CONSTRUCTION PROJECT MANAGEMENT

A Comprehensive Handbook for
Effective Site Managers Vol. 1

Steven Smith, Ph.D.

Wisdom Publishers

ISBN: 9798856095660
Imprint: Independently published

Cover design by: Art Painter
Library of Congress Control Number: 2018675309
Printed in the United States of America

To all the aspiring and current site managers, who tirelessly strive to build a better world through their exceptional dedication and leadership. Your unwavering commitment to excellence in on-site operations inspires us all. This book is dedicated to you, the visionary leaders who turn dreams into reality, construct the foundations of progress, and shape the future of our built environment.

May this handbook serve as a guiding light on your journey, empowering you with knowledge, skills, and inspiration to overcome challenges, achieve greatness, and make a lasting impact in the construction industry. Your relentless pursuit of quality, safety, and innovation is the cornerstone of progress, and we salute your unwavering spirit.

Thank you for your passion, hard work, and determination. This book is a tribute to your exceptional endeavors, and it is with profound respect and admiration that we present this to you. Your dedication to excellence fuels our mission to empower site managers worldwide, as together, we build a stronger and more sustainable tomorrow.

With heartfelt gratitude and respect,

Steven Smith, Ph.D.

Success in construction lies not only in building structures but in building strong relationships and fostering effective leadership.

JOHN C. MAXWELL

CONTENTS

INTRODUCTION

I n the dynamic and challenging world of construction, the role of a site manager is pivotal, serving as the linchpin that holds together the myriad complexities of on-site operations. Aspiring site managers, eager to step into this critical position, and seasoned professionals seeking to refine their management skills will find this comprehensive guide instrumental in mastering the art of site management.

At the heart of every successful construction project lies an effective site manager. Beyond being a mere supervisor, a site manager is a leader, strategist, and problem-solver, orchestrating a symphony of tasks, people, and resources to bring projects to fruition. The site manager's responsibilities extend far beyond the boundaries of traditional management roles, encompassing the need for technical expertise, leadership finesse, and unwavering dedication to safety, quality, and efficiency.

To excel as a site manager, a clear understanding of responsibilities and authority is paramount. This book delves into the multifaceted dimensions of the role, providing a comprehensive view of site managers' duties from pre-construction planning to project completion. Understanding the scope and limits of authority empowers site managers to make informed decisions that drive project success while ensuring adherence to legal and ethical standards.

Construction sites are dynamic environments with numerous moving parts, demanding exceptional coordination and adaptability. This guide equips site managers with the knowledge and tools to navigate these complexities effectively. From managing resources to maintaining clear communication and addressing unforeseen challenges, site managers will gain invaluable insights into streamlining operations and mitigating potential risks.

A successful site manager's profile extends beyond proficiency in one area; it requires striking a balance between leadership, technical expertise, and management skills. Effective leaders inspire and guide their teams, while technical expertise ensures accurate interpretation of architectural and engineering drawings, keeping projects on track and within specifications. Sound management skills, including budgeting, scheduling, and quality control, are pivotal in orchestrating smooth operations and delivering exceptional outcomes.

Throughout this book, readers will embark on a journey of continuous learning, honing their site management capabilities to meet the challenges of the construction landscape. Packed with practical insights, industry best practices, and time-tested techniques, this guide serves as an indispensable companion for aspiring and seasoned site managers alike.

As we explore into the world of site management, excellence becomes the standard, and effective leadership paves the way to success. Embrace the knowledge presented within these pages, and let it be the cornerstone of your journey toward becoming an exceptional site manager.

CHAPTER 1: PREPARING FOR ON-SITE OPERATIONS

I n this chapter, we will cover the essential elements of preparing for on-site operations. Effective project execution requires meticulous pre-construction planning and coordination. From conducting comprehensive site assessments and collaborating with key stakeholders to setting clear project objectives and obtaining the necessary permits, site managers will gain valuable insights to ensure a seamless and efficient start to their construction projects. By mastering the art of preparation, site managers lay the foundation for a well-organized and successful construction journey.

Pre-Construction Planning and Coordination

Pre-construction planning and coordination form the bedrock of successful on-site operations. In this crucial phase, site managers engage in a series of strategic activities, which are discussed.

Conducting Site Assessments and Feasibility Studies
In the domain of construction site management, the initial

and critical step towards successful project execution involves conducting comprehensive site assessments and feasibility studies. These early investigations lay the groundwork for informed decision-making, efficient resource allocation, and risk mitigation. For aspiring and current site managers, mastering the art of site assessments and feasibility studies is imperative to ensure a solid foundation for their construction projects.

The Significance of Site Assessments

Site assessments serve as the bedrock of project planning. As a site manager, meticulously evaluating the physical attributes of the construction site, including terrain, soil conditions, and existing structures, is vital. These assessments provide valuable insights into the site's potential challenges and advantages, enabling you to develop informed strategies for construction and resource management.

Understanding Feasibility Studies

Feasibility studies play a pivotal role in determining the viability of a construction project. As a site manager, you will conduct rigorous analyses of various factors, such as technical feasibility, economic viability, and legal considerations. These studies help you assess the project's practicality, potential risks, and estimated costs, facilitating sound decision-making before committing significant resources.

Evaluating Environmental Impact

In the contemporary construction landscape, environmental considerations hold paramount importance. Site managers must adeptly conduct environmental impact assessments to identify potential ecological risks and devise eco-friendly solutions. Adhering to sustainable practices allows you to minimize the project's ecological footprint and foster harmonious coexistence with the surrounding environment.

Engaging Stakeholders and Regulatory Authorities

During the site assessment and feasibility study phase, proactive engagement with stakeholders, including local communities, government agencies, and regulatory authorities, is essential for site managers. Effective communication and collaboration ensure that all parties' concerns and requirements are addressed, facilitating a smoother project approval process and garnering community support.

Mitigating Site-Specific Challenges

Each construction site presents its unique set of challenges. Through meticulous site assessments and feasibility studies, site managers can identify potential bottlenecks and risks, enabling proactive risk management and the formulation of contingency plans. Preparedness for unforeseen obstacles is crucial for maintaining project timelines and budgets.

Harnessing Technological Advancements

In the present era of technological advancement, site managers have an array of cutting-edge tools at their disposal for conducting site assessments and feasibility studies. Geographic Information Systems (GIS), Building Information Modeling (BIM), and drone surveys are just a few examples of technologies that enhance data accuracy and streamline decision-making processes.

Proficiency in conducting site assessments and feasibility studies is a fundamental skill for aspiring and current site managers. By diligently evaluating site conditions, understanding feasibility, and mitigating potential challenges, site managers can make informed decisions that lay the groundwork for successful on-site operations. Equipped with a thorough understanding of these essential practices, site managers will be well-prepared to navigate the complexities of construction projects and achieve excellence in their roles.

Collaborating with Project Managers and Stakeholders

In the field of construction site management, effective

collaboration with project managers and stakeholders is paramount for ensuring the success of any construction project. Aspiring and current site managers must master the art of collaboration to foster open communication, shared objectives, and harmonious teamwork throughout the project lifecycle.

The Importance of Collaboration

Collaboration forms the backbone of seamless project execution. As a site manager, forging strong collaborative relationships with project managers and stakeholders is vital for aligning efforts towards achieving project goals. By embracing a collaborative approach, site managers can optimize resource allocation, identify potential risks, and address challenges proactively.

Establishing Clear Communication Channels

Effective communication lies at the heart of successful collaboration. Site managers must establish clear and open channels of communication with project managers and stakeholders. Regular meetings, progress updates, and feedback sessions are essential for keeping all parties informed and engaged. Transparent communication fosters trust and enables prompt resolution of any emerging issues.

Aligning Objectives and Expectations

Collaboration is most fruitful when all parties share common objectives and expectations. Site managers and project managers must work together to define clear project goals, timelines, and deliverables. Additionally, understanding the needs and expectations of stakeholders, including clients, regulatory bodies, and local communities, is vital for ensuring project success.

Coordinating Resources and Tasks

Collaboration between site managers and project managers extends to resource management and task coordination.

Effective collaboration allows for seamless integration of construction activities, ensuring that resources are allocated efficiently to meet project milestones. By fostering effective coordination, site managers can optimize productivity and reduce potential bottlenecks.

Handling Challenges and Conflict Resolution

In any construction project, challenges and conflicts are inevitable. Effective collaboration equips site managers to handle these situations with tact and diplomacy. By actively engaging in conflict resolution, site managers can promote understanding and foster a spirit of teamwork among project stakeholders, ultimately enhancing project outcomes.

Engaging Stakeholders Throughout the Project

Collaboration goes beyond internal teams; it extends to engaging external stakeholders throughout the project lifecycle. Site managers must proactively involve stakeholders in decision-making processes, seek their input, and address their concerns. Engaging stakeholders fosters a sense of ownership and commitment, resulting in increased support and positive project outcomes.

Mastering the art of collaborating with project managers and stakeholders is a foundational skill for aspiring and current site managers. By embracing open communication, aligning objectives, coordinating resources, and adeptly resolving challenges, site managers can foster a culture of collaboration that elevates the success of construction projects. Effective collaboration serves as a catalyst for achieving excellence in on-site operations and solidifies site managers' position as leaders in the construction industry.

Establishing Clear Project Objectives and Deliverables

Effective construction site management hinges on the establishment of clear project objectives and deliverables. Aspiring and current site managers must master the art of defining precise project goals to provide a well-defined roadmap for their construction projects.

The Significance of Clear Project Objectives

Clear project objectives serve as the guiding principles for every construction endeavor. As site managers, articulating well-defined objectives sets the direction for the project and aligns the efforts of all team members. This clarity enables efficient resource allocation, effective risk management, and a shared sense of purpose among stakeholders. To establish clear project objectives:

- Conduct thorough discussions with project stakeholders to understand their expectations and goals for the project.
- Define specific, measurable, achievable, relevant, and time-bound (SMART) objectives to provide a focused and actionable framework.
- Ensure that objectives are feasible and aligned with the project's scope, available resources, and constraints.
- Communicate the established objectives to all team members to foster a shared understanding and commitment.

Defining Measurable Deliverables

In addition to setting clear objectives, defining measurable deliverables is essential for monitoring project progress and success. Site managers must collaborate with project managers and stakeholders to identify tangible milestones and outcomes. To define measurable deliverables:

- Break down the project into specific tasks and subtasks with clear milestones.
- Assign quantifiable metrics to each deliverable to enable accurate tracking of progress.
- Utilize project management tools and software to

monitor deliverables in real-time.

- Regularly communicate progress updates to stakeholders and team members to maintain transparency.

Ensuring Feasibility and Realism

While setting project objectives, it is imperative to ensure their feasibility and realism. As site managers, you must evaluate the project's scope, available resources, and constraints to establish attainable goals. To ensure feasibility and realism:

- Conduct comprehensive feasibility studies and assess the project's technical, financial, and logistical viability.
- Engage in constructive discussions with project stakeholders to manage expectations and avoid setting unrealistic objectives.
- Anticipate potential challenges and incorporate risk management strategies into the project planning.

Collaborative Objective Setting

Establishing clear project objectives and deliverables is a collaborative effort that involves project managers, stakeholders, and the site management team. To facilitate collaborative objective setting:

- Foster an open and inclusive environment where all stakeholders can voice their input and concerns.
- Hold regular meetings and workshops to gather feedback and ensure alignment with the project's vision.
- Emphasize the importance of teamwork and collective ownership of project goals.

SMART Objectives for Precision

To ensure the effectiveness of established objectives, site

managers often employ the SMART criteria - Specific, Measurable, Achievable, Relevant, and Time-bound.

Using the SMART framework

- Craft specific objectives that leave no room for ambiguity or misinterpretation.
- Identify measurable criteria to evaluate progress and success.
- Ensure that objectives are achievable within the project's scope and resources.
- Align objectives with the overall project vision and relevant to stakeholders' needs.
- Set time-bound objectives with clear deadlines to maintain momentum and focus.

Continual Evaluation and Adaptation

Clear project objectives are not set in stone; they require continual evaluation and adaptation. To facilitate continual evaluation and adaptation, you have to do the following:

- Regularly review project progress against established objectives.
- Be open to feedback and adjust objectives as necessary to accommodate changing project dynamics.
- Encourage a culture of learning and improvement within the project team.

Establishing clear project objectives and deliverables is a fundamental skill for aspiring and current site managers. By defining precise project goals, measuring outcomes, and ensuring feasibility, site managers lay the groundwork for successful on-site operations. Collaborative objective setting fosters a shared commitment to project success among stakeholders, while employing the SMART criteria ensures the precision and effectiveness of project objectives. Continual evaluation and adaptation enable site managers to navigate the complexities of construction projects and drive them towards successful completion.

Reviewing Construction Documents and Drawings
Thoroughly reviewing construction documents and drawings is an integral part of the preparation process for any construction project. Aspiring and current site managers must master the skill of meticulous document scrutiny to ensure precise understanding, accurate implementation, and adherence to project specifications.

The Importance of Document Review
Construction documents and drawings serve as the blueprint for the entire project. As site managers, conducting a comprehensive review is essential for grasping the project's scope, requirements, and design intent. By analyzing these documents diligently, site managers can effectively plan, coordinate, and execute the construction process. To conduct an effective document review:

- Familiarize yourself with the project scope and objectives before commencing the review.
- Develop a systematic approach to the review process, ensuring no crucial details are overlooked.
- Collaborate with project managers and relevant experts to gain insights into technical aspects.

Understanding Architectural and Engineering Drawings
Architectural and engineering drawings are the visual representation of the project's design and technical details. As a site manager, interpreting these drawings accurately is vital to translating the architect's vision into tangible structures. To understand architectural and engineering drawings:

- Familiarize yourself with common symbols, annotations, and scales used in construction drawings.
- Study floor plans, elevations, sections, and details to comprehend the overall layout and dimensions.

- Pay attention to specific construction details, such as structural elements and material specifications.

Verifying Design Intent and Specifications
During the document review, site managers must verify that the construction documents align with the project's design intent and specifications. Ensuring this alignment is crucial to avoiding discrepancies that can lead to costly rework or delays. To verify design intent and specifications:

- Cross-reference architectural and engineering drawings to confirm consistency between various plans.
- Compare the construction documents with the project's contractual requirements and scope.
- Seek clarification from designers or consultants for any ambiguities or conflicts.

Coordinating with Design Teams for Seamless Execution
Collaboration with design teams is essential to address any design-related challenges or queries that arise during the document review. Effective communication ensures that design changes, if needed, are resolved promptly, minimizing potential disruptions during the construction process. To foster effective coordination with design teams:

- Schedule regular meetings to discuss design-related matters and address any design-related issues.
- Maintain open channels of communication for design clarifications and revisions.
- Ensure that all project stakeholders are kept informed of any design updates.

Ensuring Compliance with Building Codes and Regulations
Compliance with local building codes and regulations is paramount for any construction project. Site managers must review construction documents to ensure that they meet all

applicable legal requirements and safety standards. To ensure compliance with building codes and regulations:

- Collaborate with regulatory authorities and relevant experts to verify adherence to legal requirements.
- Identify potential areas of non-compliance and develop strategies for rectification.
- Implement quality control processes to guarantee compliance throughout the project.

Reviewing construction documents and drawings is a fundamental aspect of preparing for on-site operations. By conducting meticulous document scrutiny, site managers lay the foundation for precise execution, adherence to design intent, and compliance with building codes and regulations. A thorough understanding of these essential documents empowers site managers to effectively coordinate with design teams, minimize errors, and ensure the seamless progression of construction projects.

Procuring Necessary Permits and Approvals
Securing the necessary permits and approvals is a critical phase in the preparation process of any construction project. Aspiring and current site managers must be well-versed in navigating the complex landscape of regulatory requirements to ensure compliance and a smooth start to construction.

The Importance of Permit Procurement
Obtaining permits and approvals is not just a bureaucratic formality; it is a vital legal obligation for every construction project. Site managers must recognize that permits are essential for legitimizing the project, ensuring adherence to safety standards, and preventing potential legal complications. To navigate the process of permit procurement:

- Identify the specific permits required for the project based on its scope, location, and nature of construction.

- Establish a clear timeline for acquiring permits and incorporate it into the overall project schedule.
- Appoint a designated team member responsible for managing the permit application process.

Understanding the Permitting Process

The permitting process varies based on regional regulations and the type of construction. As a site manager, you must gain a comprehensive understanding of the local permitting process to efficiently navigate through it. To understand the permitting process:

- Conduct thorough research on local building codes and zoning regulations that govern the project site.
- Engage in discussions with regulatory authorities to clarify requirements and procedures.
- Seek guidance from experienced professionals or consultants to ensure compliance with specific permit conditions.

Navigating Environmental and Regulatory Compliance

Environmental compliance is an essential aspect of the permitting process, especially for construction projects that may impact the environment. Site managers must be vigilant in assessing potential environmental impacts and mitigating adverse effects. To navigate environmental and regulatory compliance:

- Conduct environmental impact assessments to identify potential impacts on natural resources and ecosystems.
- Develop strategies to minimize environmental disturbances and ensure sustainable construction practices.
- Collaborate with environmental agencies and specialists to ensure compliance with environmental regulations.

Securing Approvals for Smooth Project Progression

Delays in obtaining permits and approvals can significantly impede project timelines and escalate costs. Site managers must proactively plan and prioritize permit procurement to maintain project momentum. To secure approvals for smooth project progression:

- Initiate the permit application process well in advance of the construction start date.
- Track the progress of permit applications and follow up with regulatory authorities as needed.
- Anticipate potential challenges or delays and devise contingency plans to avoid project interruptions.

Collaborating with Stakeholders and Authorities

Successful permit procurement requires effective collaboration with various stakeholders, including regulatory authorities, local communities, and project sponsors. Site managers must proactively engage with these stakeholders to build positive relationships and facilitate the permit approval process. To foster collaboration with stakeholders and authorities:

- Establish open communication channels to address concerns and answer inquiries promptly.
- Showcase a commitment to environmental stewardship and community welfare to gain support for the project.
- Be responsive to feedback and incorporate suggestions from stakeholders when feasible.

Procuring necessary permits and approvals is a fundamental aspect of preparing for on-site operations. Site managers play a pivotal role in navigating the permitting process, ensuring compliance with legal requirements, and securing approvals to legitimize the construction project. By understanding the intricacies of permit procurement and engaging with stakeholders and authorities effectively, site managers lay the groundwork for a successful and legally compliant construction journey.

CHAPTER 2: ORGANIZING THE CONSTRUCTION SITE

I n this section, we will explore the essential elements of organizing a construction site. Aspiring and current site managers will gain insights into strategies for efficient site layout and logistics planning, optimizing material and equipment placement, ensuring safety and accessibility, and setting up temporary facilities and infrastructure. Effective organization is vital in creating a safe, productive, and well-coordinated construction environment.

Site Layout and Logistics Planning

Site layout and logistics planning are fundamental aspects of construction site management. Our focus is on streamlining operations, optimizing workflow, and enhancing productivity to ensure you achieve a successful construction project delivery as a site manager.

Designing Efficient Site Layouts and Flow Paths

Efficient site layouts and well-planned flow paths are essential components of successful construction site management.

Aspiring and current site managers must possess the expertise to create a site layout that optimizes space, minimizes movement, and fosters a safe and productive working environment. This section will delve into the key principles and strategies for designing efficient site layouts and flow paths.

Analyzing Project Requirements and Constraints

Before embarking on the site layout design, site managers must thoroughly analyze the project's requirements, constraints, and objectives. Factors such as project scope, timeline, available space, and site conditions play a crucial role in determining the optimal layout. Engaging in discussions with project managers, stakeholders, and design teams will provide valuable insights for making informed decisions.

Zoning and Functional Segmentation

An efficient site layout begins with effective zoning and functional segmentation. Site managers must divide the construction site into designated zones, each serving specific purposes such as material storage, equipment staging, and work areas. Segmentation ensures a logical and organized site layout, minimizing unnecessary movement and congestion.

Prioritizing Safety and Accessibility

Safety should be a top priority in site layout design. Site managers must carefully plan for safe access routes, clearly marked pedestrian walkways, and separation of vehicular and foot traffic. Emergency exits and evacuation routes should be well-defined and easily accessible. Adhering to safety standards and regulations is imperative to mitigate the risk of accidents and ensure a secure working environment.

Optimal Material and Equipment Placement

Efficient site layouts optimize the placement of materials and equipment to minimize transportation distances and streamline workflow. Site managers should strategically position materials and equipment in proximity to their designated work areas, reducing handling time and increasing overall efficiency. Utilizing tools like BIM can aid in visualizing the placement of materials and equipment during the design phase.

Consideration of Future Expansion
While designing the site layout, forward-thinking site managers should consider potential future expansion needs. Anticipating changes in project requirements and incorporating flexibility in the layout allows for seamless adaptation to evolving construction demands. This approach helps to avoid disruptions and costly modifications as the project progresses.

Collaboration with Design and Construction Teams
Site managers must collaborate closely with design and construction teams throughout the layout design process. Consulting with architects, engineers, and contractors ensures that the site layout aligns with the project's design intent and construction methodologies. Regular communication facilitates adjustments and refinements based on practical insights from on-site experts.

Designing efficient site layouts and flow paths is a crucial skill for aspiring and current site managers. An effective layout not only enhances productivity but also fosters a safe and organized construction site. By carefully analyzing project requirements, prioritizing safety, optimizing material and equipment placement, and considering future expansion needs, site managers can create a well-organized and functional construction site that contributes to the overall success of the project.

Optimizing Material and Equipment Placement
Efficient material and equipment placement are crucial factors in construction site management. Aspiring and current site managers must master the art of strategic placement to minimize waste, streamline operations, and enhance

productivity. This section will delve into key principles and strategies for optimizing material and equipment placement on the construction site.

Strategic Inventory Management
Effective material placement begins with strategic inventory management. Site managers should closely collaborate with procurement teams to ensure timely delivery and accurate inventory control. Maintaining an organized storage area and implementing just-in-time inventory practices can reduce excess stock and the associated costs.

Proximity to Work Areas
Placing materials and equipment in proximity to their designated work areas can significantly impact productivity. Site managers must plan for a smooth flow of materials, minimizing transportation distances and unnecessary handling. Strategic placement reduces downtime, accelerates construction progress, and optimizes resource utilization.

Staging Areas for Efficiency
Designating staging areas for materials and equipment is a fundamental aspect of efficient construction site planning. By strategically locating these areas near high-demand work zones, site managers can facilitate seamless access and reduce delays. Properly managed staging areas minimize congestion and enhance overall site organization.

Traffic Flow and Material Handling
Site managers should carefully plan for traffic flow and material handling paths. By establishing clear and safe routes for the movement of materials and equipment, potential bottlenecks and accidents can be mitigated. Attention to the ergonomics of material handling ensures the safety and well-being of workers.

Utilization of Vertical Space

Maximizing the use of vertical space is an effective strategy to optimize material placement. Site managers can implement shelving, racks, and stacking systems to efficiently store materials without occupying excessive ground space. Utilizing vertical space also enables better organization and inventory visibility.

Equipment Accessibility and Utilization

Strategically positioning construction equipment is vital to maintaining a smooth workflow. Site managers should consider equipment accessibility for operators and ensure the proximity of equipment to the tasks they will perform. A well-planned layout reduces downtime and allows for swift equipment mobilization when needed.

Safety Considerations

Safety is paramount when optimizing material and equipment placement. Site managers must adhere to safety regulations and ensure that materials and equipment are positioned securely to prevent accidents. By eliminating potential hazards and implementing safety protocols, site managers create a safer work environment.

Optimizing material and equipment placement is an essential skill for aspiring and current site managers. Effective inventory management, proximity to work areas, strategic staging, streamlined traffic flow, vertical space utilization, equipment accessibility, and safety considerations are key components of successful optimization. By mastering these strategies, site managers can increase construction efficiency, reduce costs, and elevate overall project performance. A well-organized construction site contributes significantly to project success and enhances the reputation of site managers as proficient leaders in the construction industry.

Ensuring Safety and Accessibility on the Construction Site

Maintaining a safe and accessible construction site is of

utmost importance for aspiring and current site managers. This section explores the critical aspects of creating a secure work environment, safeguarding workers' well-being, and ensuring compliance with safety regulations.

Prioritizing Safety Culture

A strong safety culture starts with the commitment and leadership of site managers. Instilling a safety-first mindset among all team members is essential to foster a work environment where safety is a core value. By leading by example and promoting safety awareness, site managers set the tone for a culture that prioritizes the well-being of everyone on the construction site.

Identifying and Mitigating Hazards

Site managers must conduct thorough risk assessments to identify potential hazards on the construction site. From hazardous materials to uneven terrain, understanding and addressing these risks are crucial for preventing accidents and injuries. Mitigation strategies should be implemented promptly to eliminate or minimize identified hazards.

Implementing Safety Protocols and Procedures

Site managers must establish clear safety protocols and procedures tailored to the specific site and project requirements. Safety protocols should cover various scenarios, including emergency response plans, personal protective equipment (PPE) requirements, and safe work practices. Regular safety training sessions should be conducted to ensure all team members are well-versed in these protocols.

Providing Adequate Signage and Warnings

Effective communication is key to maintaining a safe construction site. Site managers must ensure the proper placement of safety signs and warnings to alert workers and visitors to potential risks. Clearly marked emergency exits, caution signs, and restricted areas contribute to a safer working environment.

Accessible Pathways and Facilities

Ensuring accessibility is vital for a construction site that promotes inclusivity and accommodates all workers. Site managers should provide accessible pathways and facilities to allow smooth movement for workers with disabilities. Wheelchair ramps, designated rest areas, and appropriate accommodations contribute to an inclusive and supportive site environment.

Regular Site Inspections and Audits

Site managers must conduct regular inspections and audits to monitor safety compliance and identify areas for improvement. These assessments should be documented, and corrective actions should be taken promptly to address any safety concerns. Collaborating with safety experts and consultants can provide valuable insights and expertise in this process.

Promoting a Reporting Culture

Site managers should encourage an open reporting culture where workers feel comfortable reporting safety hazards and incidents without fear of reprisal. Swift action should be taken in response to reported issues to rectify them and prevent recurrence.

Ensuring safety and accessibility on the construction site is a fundamental responsibility of aspiring and current site managers. By prioritizing safety culture, identifying and mitigating hazards, implementing safety protocols, providing adequate signage and warnings, ensuring accessibility, conducting regular inspections, and promoting a reporting culture, site managers create a secure and supportive work environment. A safe construction site not only protects workers from harm but also enhances productivity and contributes to the successful completion of construction projects. As leaders in the construction industry, site managers must lead by example and advocate for safety at all times.

Setting Up Temporary Facilities and Infrastructure

The establishment of temporary facilities and infrastructure is a vital aspect of construction site management. Aspiring and current site managers must possess the expertise to plan and set up these essential amenities to facilitate smooth operations and support the workforce. This section explores the key considerations and best practices for setting up temporary facilities and infrastructure on the construction site.

Temporary Office and Administrative Spaces

Creating a functional temporary office is essential for effective on-site management. Site managers must designate a centralized area for administrative tasks, project coordination, and communication. This office space should be equipped with necessary utilities, furniture, and communication systems to support seamless workflow and decision-making.

Accommodations and Welfare Facilities

Site managers must ensure the provision of proper accommodations and welfare facilities for the construction workforce. This includes comfortable sleeping quarters, sanitary facilities, and dining areas. Adequate welfare amenities contribute to worker satisfaction, productivity, and overall well-being during the construction phase.

Storage and Material Handling Facilities

Efficient storage and material handling facilities are critical for optimizing the construction process. Site managers should plan for well-organized material storage areas to prevent loss, damage, and theft. Implementing proper material handling systems, such as cranes and forklifts, helps streamline operations and reduce manual handling risks.

Security and Safety Measures

Site managers must prioritize security and safety when setting up temporary facilities and infrastructure. Installing proper lighting, surveillance systems, and access controls can help deter unauthorized access and enhance site security. Additionally, implementing safety measures, such as fire exits, first aid stations, and emergency response plans, contributes to a secure work environment.

Connectivity and Utilities

Uninterrupted connectivity and access to utilities are essential for construction site functionality. Site managers should arrange for reliable power supply, water availability, and internet connectivity to support various on-site activities. These utilities are crucial for enabling communication, data exchange, and operational efficiency.

Environmental Considerations

Site managers should be mindful of environmental considerations when setting up temporary facilities. Implementing sustainable practices, waste management systems, and erosion control measures helps minimize the environmental impact of construction activities. Additionally, adherence to local environmental regulations ensures responsible site management.

Adaptability and Scalability

Temporary facilities and infrastructure should be designed with adaptability and scalability in mind. As construction projects evolve, the site's requirements may change. Site managers should anticipate such changes and plan for facilities that can accommodate modifications and expansion as needed.

Setting up temporary facilities and infrastructure is a critical aspect of construction site management. By providing functional office spaces, welfare facilities, storage areas, security measures, and utilities, site managers create a conducive work environment for construction teams. Considerations for adaptability and environmental responsibility enhance the site's efficiency and sustainability. Effective planning and implementation of temporary facilities contribute significantly to project success and the well-being of the construction workforce. As leaders in the construction industry, site managers must demonstrate proficiency in organizing and managing temporary facilities to support seamless construction operations.

Establishing On-Site Offices and Amenities

Setting up on-site offices and amenities is a crucial aspect of construction site management. Aspiring and current site managers must ensure the provision of functional and well-equipped facilities to support effective on-site operations and provide a conducive work environment for the construction team. These on-site offices and amenities serve as the nerve center of the construction site, facilitating communication, coordination, and decision-making.

The on-site office is the command center for construction project management. It serves as a hub for site managers, project engineers, and administrative staff. This office should be strategically located to provide a clear view of the entire site, allowing managers to oversee ongoing activities and address any emerging issues promptly. Equipping the on-site office with proper communication systems, computers, and office supplies ensures efficient data exchange and administrative functions.

Accommodations and welfare facilities are crucial for the well-being and productivity of the construction workforce. Providing comfortable and hygienic sleeping quarters, sanitary facilities, and dining areas is essential for workers' physical and mental well-being. Adequate welfare amenities contribute to job satisfaction, reduce downtime, and enhance overall worker performance during the construction phase.

Storage facilities play a vital role in organizing materials and equipment on the construction site. Site managers should designate specific areas for material storage, taking into account factors such as material types, quantities, and safety considerations. A well-organized storage area minimizes material handling time, prevents loss or damage, and optimizes construction productivity.

Security measures are imperative to safeguard both personnel and valuable assets on the construction site. Implementing surveillance systems, access controls, and proper lighting helps deter unauthorized access and mitigates the risk of theft or vandalism. Additionally, establishing clear protocols for site access and visitor registration enhances overall site security.

Environmental considerations are essential when establishing on-site offices and amenities. Site managers should adopt sustainable practices to minimize the site's environmental impact. Implementing waste management systems, recycling programs, and erosion control measures demonstrates a commitment to responsible site management and environmental stewardship.

Connectivity and utilities are vital for seamless on-site operations. Reliable power supply, water availability, and internet connectivity are essential for communication, data exchange, and various construction activities. Site managers should coordinate with utility providers and ensure uninterrupted access to utilities throughout the project duration.

Adaptability and scalability are key factors to consider when establishing on-site offices and amenities. Construction projects may undergo changes or expansions, necessitating adjustments to facilities. Site managers should plan for flexibility and accommodate future modifications to ensure on-site offices and amenities can meet evolving project requirements.

Establishing on-site offices and amenities is a pivotal responsibility of aspiring and current site managers. By providing functional and well-equipped spaces for administration, accommodations, storage, security, and utilities, site managers create a conducive work environment that enhances construction efficiency and the well-being of the workforce. Environmental responsibility and adaptability in facilities design further contribute to successful construction projects. As leaders in the construction industry, site managers must demonstrate proficiency in organizing and managing on-site offices and amenities to support seamless construction operations.

Providing Essential Utilities and Services

Ensuring the provision of essential utilities and services is a critical responsibility of aspiring and current site managers. These utilities and services are the lifeblood of any construction site, facilitating seamless operations and enabling the construction team to perform their tasks efficiently. This section delves into the key utilities and services that site managers must prioritize to support the construction process effectively.

Electricity is the backbone of construction site operations. Site managers must coordinate with local utility providers to ensure a reliable power supply throughout the project duration. Properly distributed electrical outlets, power distribution boards, and backup power sources are essential to support various on-site activities, such as construction equipment operation, lighting, and office equipment.

Water supply is another crucial utility that site managers must secure. Adequate water availability is necessary for construction tasks like concrete mixing, dust suppression, and sanitation. Site managers should ensure that water sources are conveniently accessible and that sufficient water storage and distribution systems are in place.

Communication infrastructure is fundamental for effective on-site coordination and information exchange. Site managers should establish a robust communication network, including mobile phones, two-way radios, and internet connectivity, to facilitate seamless communication among team members, contractors, and stakeholders.

Waste management services are essential for maintaining a clean and organized construction site. Site managers should implement waste disposal systems, recycling programs, and regular waste collection to prevent environmental pollution and promote responsible site management.

Temporary sanitary facilities are crucial for providing adequate hygiene amenities to the construction workforce. Site managers should arrange for well-maintained portable toilets and washing stations to support workers' basic needs and maintain a healthy work environment.

Safety services and facilities are paramount for construction site well-being. First aid stations equipped with necessary medical supplies should be readily accessible in case of emergencies. Additionally, site managers should ensure that safety protocols, emergency response plans, and evacuation procedures are well-communicated to all workers.

Site managers must also plan for facilities catering to the well-being of the workforce. These amenities may include rest areas, dining facilities, and recreational spaces. Providing spaces for rest and relaxation enhances worker morale and productivity during the construction phase.

Providing essential utilities and services is a crucial aspect of construction site management. Electricity, water supply, communication infrastructure, waste management, sanitary facilities, safety services, and worker welfare amenities collectively contribute to a well-organized and efficient construction site. As aspiring and current site managers, prioritizing these utilities and services ensures the construction team's smooth workflow, safety, and overall well-being. By carefully planning and coordinating the provision of these vital resources, site managers lay the foundation for successful construction projects and foster a positive work environment for the construction workforce.

Creating a Safe and Productive Work Environment

Establishing a safe and productive work environment is a fundamental responsibility of aspiring and current site managers. A well-managed work environment not only ensures the safety and well-being of the construction workforce but also enhances overall productivity and project success. This section explores the key strategies and practices that site managers must employ to create a safe and productive work environment on the construction site.

Safety Culture and Leadership

A strong safety culture begins with the site manager's leadership and commitment to prioritizing safety above all else. Site managers must lead by example, demonstrating a dedication to safety protocols and instilling a safety-first mindset among all team members. By actively promoting safety awareness and conducting regular safety training, site managers foster a work environment where safety is ingrained in every aspect of construction operations.

Risk Identification and Mitigation

Site managers must conduct comprehensive risk assessments to identify potential hazards and risks on the construction site. Whether it's related to heavy machinery operation, working at heights, or handling hazardous materials, understanding and addressing these risks are critical for preventing accidents and injuries. Implementing robust risk mitigation strategies ensures that safety remains at the forefront of all construction activities.

Proper Equipment and Training

Providing workers with the right tools and equipment is essential for their safety and productivity. Site managers should ensure that construction equipment is well-maintained, regularly inspected, and fit for the intended tasks. Additionally, site managers must ensure that all workers receive proper training on equipment operation and safety protocols, enabling them to perform their duties with confidence and competence.

Clear Communication and Safety Protocols

Effective communication is paramount for maintaining a safe work environment. Site managers must establish clear communication channels and safety protocols to ensure that all workers are aware of potential hazards and safety procedures. Regular team meetings, toolbox talks, and safety briefings keep workers informed and engaged in maintaining a secure work environment.

Worksite Organization and Housekeeping

A well-organized construction site contributes to safety and productivity. Site managers should enforce good housekeeping practices to keep work areas clean, clutter-free, and hazard-free. Proper material storage, well-marked walkways, and clear access routes contribute to accident prevention and efficient workflow.

Incident Reporting and Investigation

Creating a culture of open reporting is vital for identifying and addressing potential safety issues. Site managers must encourage workers to report any incidents, near-misses, or safety concerns without fear of reprisal. Prompt investigation of incidents helps determine the root causes and implement corrective actions to prevent future occurrences.

Creating a safe and productive work environment is an ongoing commitment for aspiring and current site managers. By prioritizing safety culture and leadership, identifying and mitigating risks, providing proper equipment and training, establishing clear communication and safety protocols, maintaining worksite organization, and encouraging incident reporting, site managers foster a work environment where workers can perform their tasks safely and efficiently. A safe and productive work environment not only protects the well-being of the construction workforce but also contributes to the successful completion of construction projects. As leaders in the construction industry, site managers play a crucial role in ensuring that safety remains at the forefront of all construction operations.

Efficient Material Handling and Storage

Efficient material handling and storage are essential aspects of construction site management that aspiring and current site managers must master. Proper material handling ensures that materials are transported, stored, and used in a manner that minimizes waste, reduces downtime, and maximizes productivity. This section delves into the key strategies and best practices for efficient material handling and storage on the construction site.

Strategic Material Placement

Site managers must strategically plan for the placement of materials on the construction site. By arranging materials in proximity to their designated work areas, site managers reduce the time and effort required for transportation. This strategic approach optimizes material flow, minimizes worker movement, and streamlines construction operations.

Just-in-Time Delivery

Implementing a just-in-time delivery system is an effective way to reduce excess inventory and storage costs. Site managers should coordinate with suppliers to schedule material deliveries based on actual project requirements. This approach ensures that materials arrive when needed, eliminating the need for extensive on-site storage.

Organized Storage Areas
Site managers must establish organized and well-maintained storage areas to store materials efficiently. Categorizing materials and implementing a clear labeling system simplifies inventory management and enables quick identification of required materials. Proper organization minimizes material search time and optimizes inventory control.

Safe Material Handling Equipment
Investing in proper material handling equipment is crucial for a safe and efficient work environment. Site managers should provide workers with equipment such as cranes, forklifts, and conveyor belts to facilitate the movement of heavy and bulky materials. This not only enhances worker safety but also expedites construction activities.

Material Inspection and Quality Control
Site managers should conduct material inspections upon delivery to ensure that materials meet quality standards and specifications. Properly inspecting materials before use helps prevent defects, rework, and construction delays. Maintaining a vigilant approach to quality control ensures that only suitable materials are utilized in the construction process.

Just-In-Case Inventory
While the just-in-time delivery system is ideal, site managers should also maintain a small just-in-case inventory of critical materials. This backup inventory helps mitigate unforeseen delays in deliveries and ensures that construction activities can proceed without interruptions.

Minimizing Material Waste

Efficient material handling includes minimizing material waste throughout the construction process. Site managers should promote responsible usage and implement waste reduction practices. Reusing, recycling, or repurposing materials whenever possible reduces costs and supports sustainable construction practices.

Efficient material handling and storage are fundamental for the success of construction projects. By strategically placing materials, adopting just-in-time delivery, organizing storage areas, providing safe material handling equipment, conducting material inspections, maintaining a just-in-case inventory, and minimizing material waste, site managers optimize construction operations and maximize productivity. Efficient material handling reduces costs, accelerates project timelines, and enhances the overall performance of the construction site. Aspiring and current site managers must continuously refine their material handling strategies to ensure that construction projects are completed on time, within budget, and with minimal waste.

Developing Effective Material Procurement and Management Strategies

Efficient material procurement and management play a pivotal role in the successful execution of construction projects. Aspiring and current site managers must employ effective strategies to ensure a seamless flow of materials, timely deliveries, and optimal inventory control. Let's explore the key approaches and best practices to achieve efficient material procurement and management on the construction site.

Thorough Material Needs Assessment

Before embarking on any construction project, site managers should conduct a comprehensive material needs assessment. Collaborating with project engineers and stakeholders, they must identify the exact quantity, type, and quality of materials required at each stage of the project. A well-detailed material needs assessment helps prevent shortages, over-ordering, and costly delays.

Building Strong Supplier Relationships
Establishing robust and reliable supplier relationships is vital for seamless material procurement. Site managers must carefully vet potential suppliers based on their track record, capacity to meet demands, and adherence to quality standards. Building strong supplier relationships fosters trust and ensures a steady supply of materials when needed.

Streamlined Procurement Processes
Site managers should implement streamlined procurement processes to minimize bureaucracy and delays. Utilizing digital procurement systems, standardizing purchase requisitions, and optimizing approval workflows speed up the procurement cycle and enhance overall efficiency.

Effective Negotiation and Cost Management
Effective negotiation skills are invaluable for site managers during the procurement process. Negotiating favorable terms, competitive prices, and flexible payment conditions can lead to cost savings and better budget management. Site managers should continuously monitor material costs and explore cost-saving opportunities.

Optimal Inventory Control and Just-In-Time Delivery
Maintaining optimal inventory levels is crucial for efficient material management. Site managers should adopt just-in-time delivery practices, where materials are ordered and delivered precisely when needed. This minimizes excess inventory, reduces storage costs, and prevents material waste.

Rigorous Material Quality Control

Ensuring material quality is of utmost importance for construction projects. Site managers must implement rigorous quality control measures, including material inspections upon delivery and adherence to specified standards. Regular quality checks safeguard against defects, rework, and potential project delays.

Building Supply Chain Resilience

Site managers must anticipate and plan for potential supply chain disruptions. Identifying alternative suppliers and material sources helps mitigate risks associated with unforeseen circumstances, such as supplier shortages or transportation issues.

Promoting Sustainable Material Procurement

Embracing sustainable material procurement practices contributes to responsible construction and environmental conservation. Site managers should prioritize eco-friendly materials and collaborate with suppliers who adhere to sustainable practices.

Efficient material procurement and management are vital for construction project success. By conducting thorough material needs assessments, building strong supplier relationships, streamlining procurement processes, practicing effective negotiation, optimizing inventory control, prioritizing material quality control, building supply chain resilience, and embracing sustainable procurement practices, site managers ensure a smooth and productive construction process. These strategies lead to cost savings, timely project completion, and the overall success of construction projects. Aspiring and current site managers must continuously refine their procurement and management approaches to meet the evolving demands of the construction industry.

Implementing Inventory Control and Just-in-Time Delivery

Implementing effective inventory control and just-in-time (JIT) delivery practices are essential for aspiring and current site managers to optimize material management on construction sites. Proper inventory control ensures that the right materials are available when needed, minimizing excess inventory and storage costs. JIT delivery, on the other hand, ensures that materials arrive precisely when required, reducing construction delays and improving overall project efficiency. Let's explore the key aspects of implementing inventory control and JIT delivery in construction site management.

Real-Time Inventory Tracking

Site managers should utilize modern inventory tracking systems to monitor material levels in real-time. Digital tools and software allow for accurate and up-to-date inventory records, enabling managers to make informed decisions about material procurement and usage. By maintaining a clear view of inventory levels, site managers can avoid shortages and excess stockpiles.

Material Reorder Point System

Implementing a reorder point system is a crucial element of effective inventory control. Site managers must determine the minimum quantity of each material that triggers a reorder. When materials reach the predetermined reorder point, new orders are placed to ensure a continuous supply, without interruption.

Centralized Material Storage

Centralized material storage is a strategic approach to inventory control. Site managers should designate specific storage areas for materials and implement a logical organization system. This arrangement enhances material tracking and reduces the time spent searching for required materials during construction activities.

JIT Delivery Coordination

Site managers should collaborate closely with suppliers to implement JIT delivery. By sharing project schedules and material requirements, suppliers can plan deliveries accordingly, ensuring materials arrive just in time for use. JIT delivery reduces the need for extensive on-site storage and minimizes the risk of material waste.

Vendor Partnerships and Communication

Developing strong partnerships with reliable vendors is vital for successful JIT delivery. Site managers must communicate project timelines, milestones, and material needs with vendors to ensure seamless coordination. Regular communication helps resolve potential issues and promotes a shared commitment to timely deliveries.

Quality Assurance and Inspection

JIT delivery requires precise coordination, but site managers must not compromise on material quality. Implementing thorough quality assurance and inspection processes upon material delivery ensures that all materials meet specified standards. Quality control checks prevent defects and rework, avoiding costly delays in construction.

Contingency Planning

While JIT delivery optimizes material flow, site managers must have contingency plans in place for unexpected events. Unforeseen disruptions in the supply chain or project schedule may occur. A contingency plan should outline alternative suppliers and material sources to mitigate potential risks.

Implementing inventory control and just-in-time delivery practices are vital for aspiring and current site managers to excel in construction site management. Real-time inventory tracking, a material reorder point system, centralized material storage, JIT delivery coordination, vendor partnerships and communication, quality assurance, and contingency planning contribute to a streamlined material management process. By optimizing inventory levels and ensuring timely material deliveries, site managers minimize costs, reduce project timelines, and enhance construction efficiency. Embracing these strategies fosters a well-organized construction site and facilitates the successful execution of construction projects. As aspiring and current site managers, continually refining these practices is key to meeting the ever-changing demands of the construction industry.

Minimizing Material Waste and Loss through Smart Handling
Efficient material handling is not only essential for the smooth flow of construction operations but also plays a crucial role in minimizing material waste and loss. Aspiring and current site managers must adopt smart handling techniques to optimize material usage, reduce construction costs, and contribute to sustainable construction practices. This section delves into the key strategies and best practices to minimize material waste and loss through smart handling on the construction site.

Accurate Material Quantification
Precise material quantification is the foundation of smart material handling. Site managers should diligently calculate the exact quantity of materials required for each construction phase based on project specifications and design. Avoiding over-ordering and wasteful excess reduces unnecessary expenses and minimizes material waste.

Controlled Material Distribution
Site managers should implement controlled material distribution to prevent unauthorized access to materials. Unauthorized access may lead to mishandling, theft, or misuse, resulting in unnecessary material loss. Secure storage and regulated distribution ensure that materials are utilized only for designated construction activities.

Training and Awareness
Site managers must provide comprehensive training to construction personnel on proper material handling techniques. Educating workers about the importance of minimizing waste, reducing material loss, and adhering to handling guidelines encourages responsible material usage and prevents avoidable errors.

Reuse and Recycling
Encouraging the reuse and recycling of construction materials is an eco-friendly approach to minimizing waste. Site managers can explore opportunities to repurpose materials or implement recycling programs for waste generated on the construction site. This sustainable practice not only reduces material waste but also supports environmental conservation efforts.

Efficient Transportation and Handling
Efficient material transportation and handling techniques are crucial for preventing damage or spoilage during transit. Site managers should ensure that materials are transported in appropriate vehicles, stored properly during transportation, and handled with care upon arrival at the construction site.

Inventory Management and Rotation
Implementing inventory management and rotation practices ensures that older materials are used before newer ones, reducing the risk of material deterioration or obsolescence. Site managers should adopt the "first in, first out" (FIFO) method to prioritize the usage of older materials in construction activities.

Material Inspection and Quality Control
Site managers must conduct regular material inspections to identify potential defects or damages. Adhering to strict quality control measures ensures that only high-quality materials are used in construction, reducing the chances of rework or replacement due to substandard materials.

Collaboration with Suppliers

Collaborating closely with suppliers is vital for efficient material handling. Site managers should communicate specific handling requirements to suppliers to ensure that materials are delivered in optimal condition. Establishing strong relationships with suppliers fosters mutual understanding and supports the site's material handling objectives.

Minimizing material waste and loss through smart handling practices is a critical aspect of construction site management. By accurately quantifying materials, controlling their distribution, providing training and awareness, promoting reuse and recycling, implementing efficient transportation and handling, practicing inventory management and rotation, conducting material inspections, and collaborating with suppliers, site managers optimize material usage and contribute to cost-effective construction operations. These strategies not only enhance construction efficiency and reduce expenses but also align with sustainable construction principles. Aspiring and current site managers play a significant role in fostering responsible material handling practices that lead to successful construction projects and a more environmentally conscious construction industry.

CHAPTER 3: LEADING AND MANAGING THE WORKFORCE

In this chapter, we delve into the critical role of site managers as leaders and managers of the construction workforce. Effective leadership and management are key to fostering a productive and motivated team. We explore essential strategies for recruiting skilled workers, promoting teamwork and collaboration, inspiring a positive work culture, and implementing effective communication channels.

Building and Motivating a Productive Construction Team

Site managers will learn how to delegate responsibilities, empower team members, and utilize task management tools to optimize workforce efficiency. Leading and managing the workforce is a fundamental aspect of successful construction site management, and this chapter equips site managers with the necessary skills and knowledge to excel in this domain.

Recruiting and Selecting Skilled Workers

One of the most critical responsibilities of site managers is recruiting and selecting skilled workers for the construction team. A strong and capable workforce is the foundation of a successful construction project. In this section, we will explore the key strategies and best practices for attracting, identifying, and hiring skilled workers to build a productive and competent construction team.

Identifying Job Requirements
The first step in recruiting skilled workers is to clearly define the job requirements. Site managers should work closely with project engineers and stakeholders to understand the specific skills, qualifications, and experience needed for each role on the construction site. A well-defined job description will serve as a guide throughout the recruitment process.

Utilizing Multiple Recruitment Channels
To reach a diverse pool of skilled candidates, site managers should use multiple recruitment channels. This can include online job portals, industry-specific websites, social media platforms, professional networks, and collaborations with local trade schools and vocational training institutions. Widening the reach of job postings increases the chances of attracting top talent.

Leveraging Industry Networking
Site managers should actively participate in industry events, conferences, and networking opportunities. Building connections with other professionals in the construction field can lead to referrals and recommendations for potential candidates. A strong professional network can be a valuable resource in identifying skilled workers.

Conducting Rigorous Screening
During the selection process, site managers should conduct rigorous screening of candidates to assess their suitability for the role. This may include reviewing resumes, conducting

initial interviews, and administering skills assessments or practical tests. Screening ensures that only qualified and capable candidates proceed to the next stage of recruitment.

Prioritizing Soft Skills

While technical skills are essential, site managers should also prioritize soft skills when selecting candidates. Effective communication, teamwork, problem-solving abilities, adaptability, and a positive work attitude are invaluable traits that contribute to a harmonious and productive construction team.

Emphasizing Safety and Compliance

Construction sites require adherence to strict safety regulations and compliance standards. Site managers should emphasize the importance of safety during the recruitment process and look for candidates who demonstrate a commitment to safety protocols. Hiring workers who prioritize safety reduces the risk of accidents and enhances overall site efficiency.

Cultural Fit and Long-Term Potential

In addition to assessing technical skills, site managers should consider cultural fit and long-term potential when making hiring decisions. A candidate who aligns with the company's values and shows potential for growth and development within the organization is more likely to contribute positively to the team's dynamics.

Recruiting and selecting skilled workers is a crucial responsibility for site managers. By identifying job requirements, utilizing multiple recruitment channels, leveraging industry networking, conducting rigorous screening, prioritizing soft skills, emphasizing safety and compliance, and considering cultural fit and long-term potential, site managers can build a skilled and motivated construction team. A well-structured and diverse workforce lays the groundwork for successful construction projects and fosters a positive work environment that leads to overall project success. Aspiring and current site managers must continuously refine their recruitment strategies to attract top talent and build a competent and cohesive team.

Fostering Teamwork and Collaboration

Creating a strong sense of teamwork and collaboration among construction teams is fundamental to the success of any construction project. Aspiring and current site managers must actively promote a collaborative work environment, where team members work together harmoniously to achieve common goals.

To foster teamwork and collaboration, site managers should first establish clear project goals and expectations for the entire team. When team members have a shared understanding of the project's objectives, they can align their efforts and work collaboratively towards a common purpose. Clearly defined roles and responsibilities also ensure that everyone knows their contribution to the project's success.

Open and effective communication is vital for fostering collaboration. Site managers should encourage a culture of open dialogue, where team members feel comfortable sharing ideas, concerns, and feedback. Regular team meetings, huddles, and briefings provide opportunities for transparent communication and collaboration.

A positive work culture plays a significant role in building teamwork and collaboration. Site managers should lead by example and promote a supportive and respectful work environment. Recognizing and appreciating team members' contributions fosters a sense of belonging and motivates them to work collaboratively towards shared goals.

Organizing team-building activities can strengthen team bonds and improve collaboration. These activities can be both on-site and off-site and should focus on promoting trust, communication, problem-solving, and teamwork among team members.

To encourage collaboration between different disciplines or trades, site managers can form cross-functional teams. These teams bring together individuals with diverse expertise, enabling them to share knowledge, skills, and perspectives, leading to innovative problem-solving and improved efficiency.

Empowering team members to take ownership of their tasks and decisions fosters a sense of responsibility and encourages collaboration. Site managers should delegate authority appropriately and trust team members to make informed choices related to their areas of expertise.

Conflicts may arise during the course of a construction project, and site managers must address them promptly and constructively. Implementing effective conflict resolution strategies ensures that disputes are resolved amicably, without negatively impacting team dynamics.

Site managers should also facilitate regular post-project evaluations to learn from successes and failures. Recognizing successful teamwork and identifying areas for improvement fosters a culture of continuous learning and collaboration.

Fostering teamwork and collaboration among construction teams is essential for achieving project success. Building a cohesive and collaborative team contributes to the seamless execution of construction projects and ensures the achievement of project milestones within specified timelines. Aspiring and current site managers must prioritize teamwork and collaboration to elevate their construction site management skills and achieve outstanding project results.

Inspiring a Positive Work Culture and Morale

Creating a positive work culture and boosting team morale

are essential responsibilities for site managers. A positive work environment not only enhances productivity but also fosters a sense of belonging and commitment among construction teams. Aspiring and current site managers must lead by example and implement strategies to inspire a positive work culture that nurtures motivation and enthusiasm among their team members.

One of the key elements of building a positive work culture is effective leadership. Site managers should demonstrate strong leadership qualities and communicate a clear vision for the project. Leading with integrity, transparency, and empathy builds trust and confidence among the team, encouraging them to perform at their best.

Recognition and appreciation play a significant role in boosting team morale. Acknowledging the efforts and achievements of individual team members as well as the entire team creates a sense of value and appreciation. Site managers should regularly celebrate milestones and successes, publicly recognizing team members' contributions.

Moreover, fostering open and constructive communication is vital for creating a positive work culture. Site managers should encourage team members to share their ideas, concerns, and feedback. Listening attentively to their input and addressing their concerns helps build a sense of respect and inclusivity within the team.

Promoting a healthy work-life balance is another aspect that site managers can focus on. Ensuring that team members have time for rest and personal pursuits outside of work contributes to their overall well-being and job satisfaction. Encouraging occasional team outings or activities also helps strengthen team bonds and improves morale.

Training and development opportunities are essential in inspiring a positive work culture. Site managers should invest in their team's growth by providing access to skill enhancement programs and workshops. Empowering team members with new knowledge and skills not only enhances their job performance but also demonstrates the organization's commitment to their professional growth.

Lastly, maintaining a solution-oriented approach to challenges and setbacks is critical in maintaining team morale. Instead of dwelling on problems, site managers should encourage the team to focus on finding practical solutions. Demonstrating resilience and optimism in the face of challenges sets a positive example for the team.

Inspiring a positive work culture and morale is a crucial aspect of site management. Effective leadership, recognition and appreciation, open communication, work-life balance, training and development opportunities, and a solution-oriented approach are key strategies for building a positive work environment. Aspiring and current site managers must prioritize these elements to cultivate a motivated and engaged construction team that consistently delivers successful projects.

Effective Communication and Collaboration with Team Members

Effective communication and collaboration are the cornerstones of successful construction site management. Aspiring and current site managers must possess exceptional communication skills to convey project objectives, expectations, and changes clearly to the entire team. Additionally, fostering a collaborative work environment ensures that team members work together seamlessly to achieve project milestones.

First and foremost, site managers should establish clear channels of communication within the team. Regular team meetings, briefings, and updates keep everyone informed about project progress, challenges, and new developments. Providing a platform for team members to ask questions and seek clarifications fosters transparency and ensures that everyone is on the same page.

Moreover, site managers should be approachable and open to feedback from their team members. Encouraging open dialogue allows team members to express their concerns, share their insights, and offer suggestions for improvement. Listening attentively to the team's feedback demonstrates respect and value for their contributions.

Clear and concise communication is essential, especially in the fast-paced construction environment. Site managers should use simple language and avoid jargon to ensure that all team members understand the information being conveyed. Additionally, utilizing visual aids, such as charts or diagrams, can further enhance communication and understanding.

Collaboration among team members is equally crucial for project success. Site managers should actively promote teamwork and encourage cross-functional collaboration. Building a sense of camaraderie and shared purpose among team members leads to improved problem-solving and increased efficiency.

Incorporating technology and project management tools can also facilitate effective communication and collaboration. Project management software, communication platforms, and collaboration tools streamline information exchange, document sharing, and task allocation, enhancing team productivity.

Regular team-building activities and workshops further strengthen collaboration and cohesion within the team. These activities provide opportunities for team members to get to know each other on a personal level, build trust, and foster a collaborative spirit.

Finally, site managers should lead by example when it comes to communication and collaboration. Demonstrating open and respectful communication and actively participating in collaborative efforts set the tone for the entire team.

Effective communication and collaboration are essential skills for site managers to master. By establishing clear communication channels, encouraging open dialogue, using simple language, promoting collaboration, utilizing technology, organizing team-building activities, and leading by example, site managers can create a work environment that thrives on effective communication and seamless collaboration. This, in turn, leads to a motivated and engaged construction team capable of delivering successful projects.

Establishing Clear Communication Channels

Clear communication is the foundation of effective construction site management. Aspiring and current site managers must establish and maintain clear communication channels to ensure that project information is conveyed accurately and efficiently to all team members.

One of the first steps in establishing clear communication channels is to identify the different modes of communication that will be used throughout the project. This may include in-person meetings, emails, phone calls, messaging platforms, and project management software. Site managers should select communication channels based on the project's specific needs and the preferences of their team members.

Moreover, site managers should define the purpose and scope of each communication channel. For example, in-person meetings may be used for discussing project progress and addressing critical issues, while emails may be more suitable for sharing general updates and documentation. By clarifying the purpose of each channel, team members will know when and how to use them effectively.

Consistency is key to clear communication. Site managers should establish regular communication schedules to ensure that information is relayed in a timely manner. This may include daily or weekly team meetings, progress updates, and briefings. Having a structured communication plan reduces the likelihood of misunderstandings and delays in project execution.

In addition to formal communication channels, site managers should also encourage open and informal communication among team members. Building a culture of approachability and open dialogue fosters a collaborative and communicative environment. Team members should feel comfortable asking questions, seeking clarifications, and sharing feedback without hesitation.

Using visual aids and documentation can further enhance communication clarity. Site managers can use charts, diagrams, and drawings to supplement verbal communication and provide visual context for complex concepts. Written documentation, such as meeting minutes and project reports, ensures that important information is documented and easily accessible to all team members.

Lastly, site managers should be proactive in addressing communication challenges and resolving conflicts. Miscommunication or misunderstandings may occur during the project, and it is essential to address these issues promptly and constructively. Encouraging feedback from team members about communication effectiveness can also lead to continuous improvement in communication practices.

Conducting Regular Team Meetings and Briefings

Regular team meetings and briefings are vital components of effective construction site management. Aspiring and current site managers must prioritize these gatherings to ensure that the entire team stays informed, aligned, and motivated throughout the project's lifecycle.

Team meetings serve as a platform for sharing important project updates, discussing progress, and addressing any challenges or roadblocks. These meetings bring together all key stakeholders, including project managers, subcontractors, engineers, and other relevant team members. By gathering everyone in one place, site managers foster a collaborative and cohesive work environment.

To make team meetings effective, site managers should prepare an agenda in advance. The agenda should outline the topics to be discussed, the goals of the meeting, and the time allotted for each item. This helps keep the meeting focused and ensures that all necessary points are covered.

During team meetings, site managers should encourage active participation from all attendees. Team members should be encouraged to share their insights, ask questions, and offer suggestions. An inclusive meeting environment fosters a sense of ownership and engagement among team members.

Briefings are also essential for conveying critical information to the team efficiently. Unlike comprehensive team meetings, briefings are more targeted and aim to communicate specific updates or changes quickly. Site managers should conduct briefings as needed to address urgent matters and ensure that team members are promptly informed.

While team meetings and briefings are essential for information dissemination, they also serve as opportunities for team building and motivation. Site managers can use these gatherings to recognize and appreciate team members' efforts, celebrate project milestones, and boost team morale. Recognizing individual and team achievements encourages a sense of accomplishment and motivates team members to perform at their best.

Documenting key points and decisions made during team meetings and briefings is crucial. Meeting minutes should be recorded and distributed to all attendees after the meeting. This documentation ensures that everyone has access to the discussed information and provides a reference for future project planning and decision-making.

Ensuring Transparent Information Flow and Feedback Loops

Transparent information flow and feedback loops are essential elements of effective construction site management. Aspiring and current site managers must establish a communication framework that facilitates the seamless exchange of information among team members and stakeholders.

Transparency in information flow begins with establishing clear communication channels and protocols. Site managers should define how project information will be disseminated, who the key points of contact are, and how feedback and queries will be addressed. Transparency ensures that all team members have access to the necessary information and reduces the risk of miscommunication.

Regular and timely communication is a key aspect of maintaining transparent information flow. Site managers should provide regular updates on project progress, changes, and any issues that may arise. Ensuring that information is conveyed in a timely manner enables team members to make informed decisions and take appropriate actions.

In addition to sharing information, site managers must also encourage feedback from team members and stakeholders. Feedback loops allow for two-way communication, where team members can express their opinions, concerns, and suggestions. This open dialogue promotes a collaborative work environment and empowers team members to contribute to project improvements.

To foster a feedback culture, site managers should be receptive to input from their team members. Listening actively to feedback and taking it into account when making decisions demonstrates respect for their expertise and insights. Constructive feedback should be acknowledged and used to drive continuous improvement.

Implementing a formal process for feedback collection and review is crucial. Site managers can conduct regular feedback sessions or use digital platforms to gather input anonymously. Analyzing the feedback received helps identify areas that require attention and provides valuable insights for enhancing project performance.

Transparent information flow extends to external stakeholders, such as clients, subcontractors, and regulatory authorities. Site managers should ensure that relevant project information is shared with these stakeholders promptly and accurately. Transparency in communication with external parties builds trust and facilitates smoother project execution.

Site managers must also be transparent about project challenges and risks. Openly addressing potential issues allows the team to prepare and implement mitigation strategies. Transparent risk communication ensures that all stakeholders are aware of potential impacts on the project's timeline and budget.

Delegating Responsibilities and Task Management

Delegating responsibilities and effective task management are critical skills that aspiring and current site managers must master. Construction projects are complex and involve numerous tasks that require attention to detail and efficient execution. Delegating tasks to the right individuals and managing them effectively are key to ensuring the project progresses smoothly and meets its objectives.

One of the first steps in delegating responsibilities is to assess the strengths and capabilities of each team member. Site managers should have a clear understanding of their team's skills, expertise, and experience. By matching tasks to the appropriate team members, site managers can ensure that tasks are completed efficiently and to a high standard.

Delegation is not about offloading work; it is about empowering team members to take ownership of their assigned tasks. Site managers should communicate the importance of each task and its contribution to the overall project. By providing context and purpose, team members are more likely to approach their responsibilities with commitment and enthusiasm.

Effective task management requires setting clear objectives and deadlines for each task. Site managers should communicate the expectations and deliverables to team members, along with any relevant guidelines or specifications. Regularly reviewing progress and providing feedback helps keep tasks on track and ensures that any issues are addressed promptly.

While delegating responsibilities, it is also essential to provide adequate support and resources to the team. Site managers should be available to answer questions, offer guidance, and provide assistance when needed. Clear communication channels should be established to facilitate ongoing discussions and updates.

Additionally, site managers must trust their team members to handle their assigned tasks. Micromanaging can lead to reduced productivity and morale among team members. Trusting the expertise and judgment of the team empowers them to take initiative and make informed decisions.

Monitoring task progress is a crucial aspect of effective task management. Site managers can use project management software or other tools to track task completion, identify bottlenecks, and manage resources efficiently. Regular progress reports and meetings allow site managers to stay informed and address any issues that may arise promptly.

Assigning Roles and Responsibilities to Maximize Efficiency
Assigning roles and responsibilities is a crucial aspect of construction site management that aspiring and current site

managers must excel at. The success of a construction project heavily relies on how efficiently tasks are distributed among team members. By strategically assigning roles and responsibilities, site managers can ensure that the right people are in the right positions, maximizing efficiency and productivity.

The first step in assigning roles is to analyze the project's scope and requirements. Site managers should have a comprehensive understanding of the project's goals, deliverables, and timeline. This information allows them to determine the specific roles needed to accomplish each aspect of the project successfully.

As site managers identify roles, they should also consider the skill sets and expertise of their team members. Each team member possesses unique talents, experiences, and qualifications. Assigning roles that align with their strengths can lead to increased job satisfaction and better outcomes for the project.

To ensure that roles and responsibilities are clear, site managers should communicate the expectations and deliverables associated with each role. A well-defined role description outlines the key responsibilities, authority, and accountability of each team member. This clarity reduces the risk of misunderstandings and promotes a cohesive and organized work environment.

While assigning roles, site managers should foster a collaborative approach. Involving team members in the role assignment process can increase their sense of ownership and commitment to their responsibilities. Moreover, seeking input from team members about their preferences and aspirations can help align their interests with the project's needs.

Efficiently assigning roles also involves considering workload distribution. Site managers should distribute tasks fairly and avoid overburdening any individual team member. Balancing the workload ensures that each team member can dedicate sufficient time and effort to their responsibilities without compromising quality or efficiency.

Flexibility is essential in role assignment, especially in dynamic construction environments. As project requirements change, site managers should be prepared to adjust roles and responsibilities accordingly. This adaptability enables the team to respond to unforeseen challenges and opportunities effectively.

Assigning roles and responsibilities should not be a one-time event. Regularly evaluating the performance and progress of team members helps site managers identify any skill gaps or training needs. Offering professional development opportunities can further enhance the team's capabilities and overall efficiency.

Delegating Authority and Empowering Team Members

Delegating authority and empowering team members are essential aspects of effective site management. Aspiring and current site managers must recognize that they cannot handle every decision and task on their own. Delegating authority to the right individuals and empowering them with responsibility fosters a sense of ownership, collaboration, and initiative within the construction team.

Delegating authority involves entrusting team members with decision-making power within their assigned roles. Site managers should identify team members who have the necessary skills, expertise, and judgment to handle specific tasks and challenges. By delegating authority, site managers not only relieve their workload but also acknowledge the competence and expertise of their team members.

An important aspect of delegating authority is setting clear boundaries and expectations. Site managers should communicate the scope of authority and the level of decision-making that team members have. This ensures that team members understand their responsibilities and are aware of any limitations.

Empowering team members goes beyond simply assigning tasks; it involves encouraging them to take ownership and initiative in their work. Site managers should create a supportive environment that allows team members to voice their ideas, suggest improvements, and take calculated risks. Empowered team members are more likely to contribute creatively and proactively to the project's success.

To foster empowerment, site managers should provide the necessary resources, support, and training to help team members grow in their roles. Offering mentorship and professional development opportunities allows team members to enhance their skills and confidence, further increasing their effectiveness in decision-making.

Site managers should actively listen to and value the input of team members. Encouraging open communication and feedback ensures that team members feel heard and respected. Site managers should consider team members' perspectives when making decisions, fostering a collaborative work environment.

While delegating authority and empowering team members, site managers should also be available for guidance and support. Team members may encounter challenges or require assistance in their roles, and site managers should be ready to offer help or guidance whenever needed.

By delegating authority and empowering team members, site managers create a dynamic and self-sufficient construction team. Empowered team members take ownership of their responsibilities, leading to increased motivation and productivity. They are more likely to seek solutions independently and contribute to the overall success of the project.

Implementing Task Management Tools and Techniques

Effective task management is a fundamental aspect of construction site management. As construction projects involve a multitude of tasks and activities, aspiring and current site managers must leverage task management tools and techniques to ensure smooth execution, timely completion, and overall project success.

Task management tools play a crucial role in organizing and tracking tasks throughout the project lifecycle. These tools offer features such as task assignment, progress tracking, deadline setting, and collaboration capabilities. Site managers can choose from a variety of digital task management platforms or project management software to streamline their workflow and keep the team on track.

With the help of task management tools, site managers can assign specific tasks to individual team members and define clear deadlines for completion. Team members can access the platform to view their assigned tasks, monitor progress, and update their status in real-time. This level of transparency ensures that everyone on the team is aware of their responsibilities and the project's current status.

Moreover, task management tools facilitate collaboration and communication among team members. They allow for easy sharing of information, documents, and updates, enabling seamless collaboration and reducing the risk of miscommunication. Team members can exchange feedback and comments directly within the platform, fostering efficient communication and problem-solving.

Task management techniques complement the use of tools by providing a structured approach to task prioritization and execution. One such technique is the use of task prioritization methods, such as the Eisenhower Matrix or the Critical Path

Method. These techniques help site managers identify critical tasks and prioritize them based on their urgency and impact on project milestones.

Breaking down complex tasks into smaller, manageable subtasks is another effective task management technique. Site managers can use work breakdown structures to divide large tasks into smaller components, making it easier to track progress and allocate resources accordingly.

Regular task review meetings are vital for effective task management. Site managers can conduct meetings with team members to discuss task progress, address challenges, and identify any adjustments needed in the project plan. These meetings provide an opportunity to realign priorities, adjust timelines, and ensure that the project stays on course.

Site managers should also foster a culture of accountability within the team. Each team member should take ownership of their assigned tasks and be accountable for meeting deadlines and deliverables. Encouraging a sense of responsibility fosters a high level of commitment and dedication among team members.

Finally, site managers should continuously evaluate and refine their task management processes. Lessons learned from previous projects can inform improvements and optimizations in future task management endeavors. By learning from past experiences and seeking feedback from team members, site managers can enhance their task management strategies over time.

CHAPTER 4:
ENSURING ON-
SITE SAFETY AND
COMPLIANCE

S ite managers have a vital role in ensuring on-site safety and compliance. They prioritize safety, adhere to regulations, involve the workforce, and foster a positive work environment. This promotes productivity and project success.

Developing a Comprehensive
Site Safety Plan

Developing a comprehensive site safety plan is a critical step in ensuring on-site safety and compliance. The site safety plan serves as a roadmap for identifying potential hazards, assessing risks, and implementing safety measures throughout the construction project.

Site managers, in collaboration with safety experts and team members, craft the site safety plan to address specific project requirements and potential challenges. The plan outlines the safety procedures, protocols, and emergency response strategies that will be implemented to safeguard the construction workforce and visitors.

The first phase of developing the site safety plan involves conducting a thorough risk assessment. Site managers identify potential hazards and evaluate the level of risk associated with each activity. This includes assessing risks related to equipment operation, working at heights, hazardous materials, and other construction-specific dangers.

Based on the risk assessment, the site safety plan outlines safety measures and preventive actions to mitigate identified risks. It includes guidelines for using personal protective equipment (PPE), setting up safety barriers, and implementing safe work practices.

Moreover, the site safety plan establishes communication channels for disseminating safety information to the workforce. It ensures that team members are aware of safety protocols and are equipped with the knowledge to respond to emergencies effectively.

The plan also includes provisions for regular safety training and drills. Site managers schedule training sessions to educate team members about potential hazards and best practices. Emergency drills are conducted to practice evacuation procedures and emergency responses.

Compliance with safety regulations and industry standards is an integral part of the site safety plan. Site managers ensure that the plan aligns with local, state, and federal safety requirements. Regular audits and inspections are conducted to confirm adherence to the safety plan and compliance with regulations.

The site safety plan is a living document that evolves with the project's progression. As the project advances, site managers continuously review and update the plan to accommodate changes in scope, risk assessments, or new safety requirements.

By developing a comprehensive site safety plan, site managers proactively promote a safety-oriented work culture, reduce accidents, and enhance overall project efficiency. The plan serves as a guide to protect the well-being of the construction workforce and lays the foundation for a successful construction project.

Identifying Site-Specific Hazards and Risk Mitigation Strategies

One of the primary responsibilities of site managers is to identify site-specific hazards and develop risk mitigation strategies to ensure a safe working environment for the construction workforce. Site-specific hazards refer to potential dangers that are unique to a particular construction site, and they require careful assessment and proactive measures to prevent accidents and injuries.

The process of identifying site-specific hazards begins with a comprehensive site assessment. Site managers, along with safety experts and relevant stakeholders, conduct a thorough inspection of the construction site to identify potential hazards. This assessment encompasses various aspects, including the surrounding environment, existing structures, terrain, and weather conditions.

Common site-specific hazards may include uneven terrain, unstable soil, confined spaces, proximity to traffic, overhead power lines, and adverse weather conditions. Identifying these hazards is crucial as they can pose significant risks to the workforce and project execution if not appropriately managed.

Once the hazards are identified, site managers proceed to develop risk mitigation strategies. These strategies aim to reduce or eliminate the identified risks, ensuring the safety of workers and visitors on the site.

Risk mitigation strategies often involve implementing engineering controls, administrative controls, and personal protective equipment (PPE) measures. Engineering controls focus on modifying the physical environment to minimize hazards. For instance, installing guardrails to prevent falls from heights or using proper ventilation systems to control exposure to harmful substances.

Administrative controls involve establishing protocols and procedures to guide safe work practices. This includes implementing clear signage, creating safe work zones, and developing emergency response plans. Regular safety training and toolbox talks also fall under administrative controls to educate the workforce about site-specific hazards and safety protocols.

In cases where hazards cannot be entirely eliminated, PPE becomes essential in protecting workers. Site managers ensure that the workforce is provided with appropriate PPE, such as hard hats, safety goggles, gloves, and high-visibility vests, based on the specific risks present on the site.

Moreover, communication and collaboration are integral to the success of risk mitigation strategies. Site managers should actively engage with the workforce, seeking input and feedback on safety concerns and potential hazards they may encounter during their tasks. This open communication fosters a safety-oriented work culture where everyone actively contributes to identifying and mitigating risks.

Regular safety audits and inspections are vital for continually assessing the effectiveness of risk mitigation strategies and making necessary adjustments as the project progresses. Site managers conduct these evaluations to identify any new hazards that may emerge during construction and to confirm that existing mitigation measures remain effective.

Implementing Safety Policies and Procedures

Implementing robust safety policies and procedures is a fundamental aspect of effective site management. Safety policies and procedures serve as a framework to guide the workforce in adopting safe work practices, adhering to regulatory requirements, and preventing accidents on the construction site.

Safety policies are overarching statements that outline the site's commitment to maintaining a safe and healthy work environment. They set the tone for safety expectations and highlight the site's dedication to safeguarding the well-being of all personnel involved in the project. Safety policies should be communicated clearly to all team members, and their adherence should be a non-negotiable aspect of the construction process.

Safety procedures, on the other hand, are detailed guidelines that specify how specific tasks should be performed safely. They provide step-by-step instructions on how to handle equipment, use tools, and execute construction activities in a manner that minimizes risk. Site managers work in collaboration with safety experts to develop these procedures, ensuring that they align with industry best practices and comply with relevant regulations.

The implementation of safety policies and procedures begins with thorough training for all workers on the construction site. Site managers conduct safety orientations for new team members, ensuring that they understand the safety policies and procedures before commencing work. Additionally, refresher training sessions are organized regularly to reinforce safety knowledge and promote a culture of continuous improvement.

To maximize the effectiveness of safety policies and procedures, site managers actively involve the workforce in their development. Workers often possess valuable insights into the specific hazards they encounter on the job. By seeking input from team members, site managers can tailor safety procedures to address site-specific risks more effectively.

Regular safety meetings and toolbox talks are essential components of the implementation process. These gatherings provide an opportunity to discuss safety-related topics, address any emerging safety concerns, and share lessons learned from past incidents. Safety meetings encourage open communication between management and workers and foster a collaborative approach to safety on the construction site.

Enforcing safety policies and procedures requires consistent monitoring and evaluation. Site managers conduct regular safety inspections and audits to assess the site's adherence to established safety guidelines. Any identified deficiencies are promptly addressed, and corrective actions are implemented to enhance safety protocols.

Site managers should lead by example in adhering to safety policies and procedures. Demonstrating a strong commitment to safety fosters a safety-conscious work culture among the team. When workers observe site managers prioritizing safety, they are more likely to do the same, and safety becomes ingrained in the construction process.

Conducting Safety Training and Education Programs

Safety training and education are integral components of effective site management, ensuring that the construction workforce is equipped with the knowledge and skills to work safely on the site. Aspiring and current site managers must prioritize safety training to instill a safety-conscious work culture and reduce the likelihood of accidents and injuries.

Safety training begins with comprehensive safety orientations for all new team members joining the construction site. During these orientations, site managers introduce workers to the site's safety policies, procedures, and emergency response protocols. They also familiarize new team members with the specific hazards they may encounter on the site and the appropriate measures to mitigate these risks.

Site managers collaborate with safety experts and trainers to develop tailored safety training programs that address the unique hazards present on the construction site. These programs cover a wide range of safety topics, such as fall protection, hazardous material handling, electrical safety, and equipment operation. The training content aligns with industry best practices and regulatory requirements.

Continuous safety education is equally important. Site managers organize regular toolbox talks and safety meetings to reinforce safety knowledge and provide updates on safety-related issues. These short, focused sessions allow site managers to address emerging safety concerns, share lessons learned from past incidents, and discuss best practices for mitigating risks.

Moreover, safety education extends beyond on-site training sessions. Site managers encourage team members to pursue additional safety certifications and training courses to enhance their safety expertise. These certifications may include OSHA (Occupational Safety and Health Administration) training, First Aid/CPR training, and specialized safety certifications relevant to the construction industry.

Incorporating hands-on training and practical exercises is vital in ensuring that workers understand and can apply safety concepts effectively. Site managers organize mock drills and emergency simulations to prepare the workforce for real-life safety scenarios. These exercises enable workers to practice their response to emergencies, reinforcing their ability to act swiftly and confidently in critical situations.

Site managers also leverage technology to enhance safety training and education. Virtual reality (VR) simulations and interactive training modules offer immersive experiences that simulate on-site hazards and allow workers to practice safety protocols in a controlled environment. Such training approaches are highly effective in engaging the workforce and improving retention of safety knowledge.

Tracking and documenting safety training and education efforts are essential for ensuring accountability and compliance. Site managers maintain comprehensive records of all safety training sessions, certifications obtained, and toolbox talk topics covered. These records provide evidence of the site's commitment to safety and assist in audits or inspections.

Implementing Safety Regulations and Compliance Measures
Aspiring and current site managers play a pivotal role in ensuring that safety regulations and compliance measures are diligently adhered to on the construction site. Safety regulations are established by government agencies and industry bodies to safeguard the health and well-being of construction workers and the general public. Compliance with these regulations is not only a legal obligation but also a moral imperative to prevent accidents and create a safe work environment.

Site managers must first familiarize themselves with the relevant safety regulations and standards applicable to their specific construction project. These regulations may include OSHA (Occupational Safety and Health Administration) standards, local building codes, environmental protection requirements, and specific industry guidelines. Staying up-to-date with any changes or updates to these regulations is crucial in maintaining compliance.

Implementing safety regulations starts with effective communication to the entire workforce. Site managers must clearly convey the importance of adhering to safety regulations and the potential consequences of non-compliance. Workers should understand that safety is a shared responsibility, and each individual's commitment to following the rules contributes to a safer work environment for everyone.

To ensure compliance, site managers conduct regular inspections and audits to monitor the site's adherence to safety regulations. These inspections assess whether safety measures, such as protective barriers, signage, and personal protective equipment (PPE), are properly implemented. Any deficiencies or non-compliance identified during inspections are promptly addressed and corrected.

Moreover, site managers work collaboratively with safety officers and designated safety representatives to ensure that all safety protocols are being followed diligently. These safety officers help reinforce safety training, provide guidance to workers, and report any concerns to site management for immediate action.

To maintain compliance, site managers create a culture of accountability and transparency. Workers are encouraged to report any safety issues or near-miss incidents without fear of reprisal. By fostering open communication, site managers can address potential hazards proactively and prevent accidents before they occur.

In the event of a safety violation or non-compliance, site managers take swift and appropriate corrective actions. This may include retraining, disciplinary measures, or the implementation of additional safety measures. The goal is to instill a strong message that safety violations are not tolerated and that the health and safety of the workforce are paramount.

Site managers should also be proactive in engaging with relevant regulatory authorities. Building positive relationships with inspectors and regulators can facilitate smoother inspections and compliance checks. Additionally, proactive engagement shows a commitment to safety and a willingness to work collaboratively with regulatory agencies to uphold safety standards.

Adhering to Occupational Health and Safety Standards

Occupational Health and Safety (OHS) standards are a set of guidelines and regulations designed to protect the health, safety, and well-being of workers on the construction site. Aspiring and current site managers have a crucial responsibility to ensure strict adherence to these standards to prevent accidents, injuries, and occupational hazards.

Site managers must first familiarize themselves with the specific OHS standards applicable to their construction project. These standards are typically set by government agencies and may vary based on the location, nature of the project, and the type of construction activities involved. Compliance with OHS standards is not only a legal requirement but also a moral obligation to prioritize the health and safety of the workforce.

Implementing OHS standards starts with creating a safety-conscious work culture. Site managers lead by example, demonstrating a commitment to safety through their actions and decisions. They communicate the significance of adhering to OHS standards to all team members and promote a shared responsibility for maintaining a safe work environment.

Site managers conduct regular safety inspections to identify potential hazards and ensure that safety measures are in place. These inspections encompass the assessment of working conditions, equipment, and procedures to verify compliance with OHS standards. Any deviations or safety issues identified during inspections are promptly addressed and rectified.

To foster a safe work environment, site managers prioritize safety training and education. Workers receive comprehensive training on OHS standards, safe work practices, and the proper use of personal protective equipment (PPE). Training sessions are conducted for all team members, including subcontractors, to ensure a consistent understanding of safety protocols.

Effective communication is essential in adhering to OHS standards. Site managers establish clear communication channels for reporting safety concerns, near-miss incidents, and potential hazards. Encouraging open communication empowers workers to voice their safety-related observations, enabling site managers to take proactive measures to address any safety issues.

Incorporating regular safety meetings and toolbox talks further reinforces the importance of OHS standards. These gatherings serve as platforms to discuss safety updates, share best practices, and address safety challenges on the construction site. Safety meetings facilitate open dialogue between management and workers and foster a collaborative approach to safety.

Site managers also work closely with safety officers and designated safety representatives to ensure ongoing compliance with OHS standards. These safety officers assist in the implementation of safety protocols, monitor adherence to OHS standards, and provide guidance to workers when needed.

Continuous improvement is an integral part of adhering to OHS standards. Site managers analyze safety data, incident reports, and feedback from workers to identify opportunities for enhancement. Regular evaluation and refinement of safety measures help create a safer work environment and promote a culture of continuous improvement.

Enforcing Safety Protocols and Rules
Enforcing safety protocols and rules is a critical aspect of effective site management to ensure the well-being of the construction workforce and prevent accidents or injuries on the construction site. Aspiring and current site managers play a central role in establishing a strong safety culture by upholding safety protocols and rules with unwavering commitment.

Site managers must communicate safety protocols clearly

and effectively to all workers, subcontractors, and visitors on the construction site. They emphasize the importance of compliance with safety rules and the potential consequences of non-adherence. It is essential to create an environment where safety is not just a priority but a core value embraced by everyone involved in the project.

Consistency is key in enforcing safety protocols. Site managers ensure that safety rules are consistently applied across the site and that no exceptions are made. By holding everyone to the same standard, site managers demonstrate their dedication to the safety of the entire workforce and create a sense of fairness and accountability.

Leading by example is a powerful way to enforce safety protocols. When site managers visibly adhere to safety rules and actively participate in safety measures, they inspire others to follow suit. Demonstrating a personal commitment to safety reinforces the message that safety is non-negotiable and sets a positive example for the entire team.

Site managers conduct regular safety inspections to monitor and evaluate the implementation of safety protocols. These inspections involve examining the use of personal protective equipment (PPE), the condition of safety barriers, and the adherence to safety guidelines during various construction activities. Any observed safety violations are addressed immediately, and corrective actions are taken to prevent recurrence.

Enforcement of safety protocols also involves providing timely and constructive feedback. Site managers offer praise and recognition to individuals and teams who consistently follow safety rules. Similarly, when safety violations occur, site managers address the issue promptly and provide necessary guidance to ensure corrective action is taken.

Effective communication plays a vital role in enforcing

safety protocols. Site managers maintain open channels of communication with the entire workforce, encouraging workers to report any safety concerns or potential hazards. This fosters a culture where workers feel comfortable raising safety-related issues, enabling site managers to address them proactively.

When necessary, site managers implement disciplinary actions for repeated or severe safety violations. These actions serve as a deterrent against unsafe behavior and reinforce the seriousness of safety compliance. However, the focus remains on education and prevention, and disciplinary measures are used as a last resort.

Conducting Regular Safety Audits and Inspections

Regular safety audits and inspections are essential components of effective site management to ensure a safe and hazard-free construction site. Aspiring and current site managers must proactively conduct these audits to identify potential safety hazards, evaluate the effectiveness of safety measures, and implement corrective actions to maintain a high standard of safety.

Safety audits involve a comprehensive and systematic review of the construction site's safety practices, procedures, and conditions. Site managers, along with safety officers or designated representatives, conduct these audits at regular intervals throughout the project's duration. The goal is to assess the site's adherence to safety protocols, identify any deficiencies, and implement improvements.

During safety audits, site managers carefully examine various aspects of the construction site, including the proper use of personal protective equipment (PPE) by workers, the availability and condition of safety barriers and signage, the functionality of safety equipment, and the organization of work areas to prevent potential hazards.

Inspections focus on specific areas or activities within the construction site to ensure compliance with safety standards. Site managers may conduct inspections during critical phases of the project, such as when heavy machinery is in use, during concrete pouring, or when working at heights. By targeting these specific activities, site managers can address potential risks associated with them.

Documentation is a crucial aspect of safety audits and inspections. Site managers maintain detailed records of the audit findings, inspection results, and any corrective actions taken. This documentation serves as a reference for future audits and allows site managers to track progress in addressing safety concerns.

A key benefit of safety audits and inspections is the opportunity to engage with workers and subcontractors. Site managers use these occasions to communicate the importance of safety and to reinforce safety training and procedures. Workers are encouraged to provide feedback on safety issues, fostering a collaborative approach to safety improvement.

If safety violations or non-compliance are identified during audits or inspections, site managers take immediate action to rectify the issues. This may involve retraining workers, implementing additional safety measures, or addressing gaps in safety protocols. The goal is to address safety concerns proactively and prevent potential accidents.

Moreover, site managers also use safety audits and inspections to identify areas of excellence. Recognizing and rewarding teams that consistently demonstrate a commitment to safety reinforces the significance of safety compliance and motivates others to follow suit.

As the construction site evolves, safety audits and inspections are regularly revisited to adapt to changing conditions and to ensure ongoing compliance with safety standards. Continuous improvement is at the core of these processes, and site managers are dedicated to creating a safe and secure work environment for everyone involved in the project.

Conducting Regular Safety Inspections and Training

Safety is a paramount concern on any construction site, and conducting regular safety inspections and training is an integral part of effective site management. Aspiring and current site managers take proactive measures to ensure a safe work environment by conducting systematic safety inspections and providing comprehensive safety training for all workers.

Safety inspections are scheduled at regular intervals throughout the construction project's duration. These inspections involve a thorough examination of the construction site to identify potential hazards, assess the effectiveness of safety measures, and ensure compliance with safety regulations and protocols. Site managers, safety officers, and designated representatives collaborate to conduct these inspections.

During safety inspections, site managers carefully observe the proper use of personal protective equipment (PPE) by workers, the condition and functionality of safety equipment, the organization of work areas to prevent tripping and slipping hazards, and the adherence to safety signage and barriers. Any identified safety concerns or violations are documented, and appropriate corrective actions are promptly implemented.

Moreover, safety inspections provide an opportunity for site managers to engage with workers, subcontractors, and other stakeholders. Site managers use these interactions to communicate the significance of safety, encourage reporting of safety concerns, and promote a collaborative approach to safety improvement. Workers are encouraged to ask questions and provide feedback during inspections, fostering an open and proactive safety culture.

In addition to inspections, site managers prioritize safety training for all workers. Comprehensive safety training programs are conducted at the onset of the project and periodically throughout the construction process. New workers receive thorough onboarding training, while existing workers participate in refresher courses to reinforce safety protocols.

Safety training covers a range of topics, including the proper use of PPE, safe work practices for specific tasks, emergency response procedures, and hazard recognition and mitigation. Site managers ensure that all workers are well-versed in the safety guidelines and understand their roles and responsibilities in maintaining a safe work environment.

Site managers leverage safety training as an opportunity to educate workers on industry best practices and the latest safety standards. Workers are encouraged to share their experiences and insights, fostering a culture of continuous learning and improvement.

Site managers also collaborate with safety experts and industry professionals to deliver specialized training sessions as needed. For instance, when specific construction activities or equipment require unique safety considerations, site managers ensure that workers receive targeted training to safely carry out those tasks.

By conducting regular safety inspections and providing ongoing safety training, site managers demonstrate their commitment to creating a culture of safety excellence on the construction site. These proactive measures not only prevent accidents and injuries but also enhance productivity and build trust among the construction team.

Inspecting Construction Work Areas and Equipment
Inspecting construction work areas and equipment is a critical aspect of ensuring safety, quality, and efficiency on the construction site. Aspiring and current site managers are

diligent in conducting regular inspections to identify potential hazards, assess the condition of equipment, and maintain a high standard of workmanship.

Work Area Inspections

Site managers conduct comprehensive inspections of work areas to verify that they are organized, clean, and free from potential hazards. This includes ensuring that walkways are clear of obstructions, materials are properly stored, and waste is promptly removed. Site managers also inspect scaffolding, ladders, and other structures to confirm their stability and adherence to safety regulations.

During work area inspections, site managers assess compliance with safety protocols and proper use of personal protective equipment (PPE) by workers. They also verify that safety signage and barriers are in place to warn of potential dangers and protect workers and visitors from entering restricted zones.

Additionally, work area inspections provide an opportunity to monitor the progress of construction activities, identify potential delays, and assess resource allocation to ensure optimal productivity.

Equipment Inspections

Regular inspections of construction equipment are essential to ensure their safe operation and prevent unexpected breakdowns. Site managers work closely with equipment operators and maintenance teams to conduct thorough equipment inspections.

During equipment inspections, site managers verify that all machinery is in good working condition, with no leaks, malfunctions, or worn-out components. They also check that safety features and emergency shut-off mechanisms are functional and accessible.

Site managers ensure that equipment operators are trained in proper equipment operation and maintenance to minimize the risk of accidents and prolong the lifespan of the machinery. They also collaborate with maintenance teams to schedule regular servicing and address any identified issues promptly.

Documentation and Reporting

Detailed documentation is crucial for work area and equipment inspections. Site managers maintain comprehensive records of inspection findings, including any safety concerns, equipment maintenance schedules, and corrective actions taken. This documentation serves as a reference for future inspections and helps site managers track the progress of identified issues.

In the event that safety hazards or equipment malfunctions are discovered during inspections, site managers promptly address the issues and implement corrective measures. They collaborate with relevant teams to ensure that necessary repairs, adjustments, or safety enhancements are made without delay.

Continuous Improvement

Work area and equipment inspections are not merely routine tasks but opportunities for continuous improvement. Site managers use the insights gained from inspections to identify trends, address recurring issues, and implement preventive measures. They encourage open communication among the construction team to promote a culture of safety and efficiency.

Inspecting construction work areas and equipment is an indispensable aspect of site management to uphold safety standards and maintain operational efficiency. Through meticulous inspections, prompt actions, and comprehensive documentation, site managers demonstrate their commitment to creating a safe and productive work environment for the construction team. Work area and equipment inspections are an integral part of responsible site management, ensuring that construction projects progress smoothly, on schedule, and with the utmost regard for the well-being of all involved.

Addressing Safety Violations and Non-Compliance

Aspiring and current site managers understand the critical importance of addressing safety violations and non-compliance on the construction site. Promptly addressing these issues is essential to maintain a safe work environment and prevent potential accidents or injuries. Site managers take a proactive and assertive approach in dealing with safety violations, emphasizing the significance of adherence to safety protocols.

Identification of Safety Violations

Site managers conduct regular safety inspections and encourage workers to report any safety concerns they observe. This collaborative approach enables the timely identification of safety violations or non-compliance with established safety standards. Site managers also engage with workers and subcontractors to ensure that all individuals on the site are vigilant in recognizing potential hazards.

Taking Immediate Action

When safety violations are identified, site managers take immediate action to rectify the situation. The safety of the construction team is the top priority, and no compromises are made in addressing safety concerns. Site managers intervene directly to halt unsafe activities and implement corrective actions to eliminate or mitigate the identified hazards.

Corrective Measures and Education

Site managers work closely with workers involved in safety violations to ensure that they understand the risks and implications of their actions. They provide on-the-spot education and training to reinforce proper safety practices and protocols. Workers are encouraged to ask questions and seek clarification to avoid repeating the same mistakes in the future.

In addition to immediate corrective measures, site managers take a proactive approach to prevent similar violations from occurring. They conduct targeted safety training sessions, addressing specific areas of concern and reinforcing the importance of adhering to safety guidelines.

Enforcing Disciplinary Actions

Addressing safety violations sometimes requires enforcing disciplinary actions, especially for repeated or severe infractions. While the primary goal is to educate and improve safety awareness, site managers recognize that enforcing consequences for non-compliance is necessary to uphold safety standards and instill a culture of responsibility.

Disciplinary actions may include verbal warnings, written reprimands, additional safety training, or, in extreme cases, removal from the construction site. The intent is not punitive but rather to emphasize the gravity of safety violations and to ensure that all workers prioritize safety above all else.

Continuous Improvement and Collaboration

Site managers view the process of addressing safety violations as an opportunity for continuous improvement. They engage in open communication with workers and subcontractors to understand the root causes of non-compliance and work collaboratively to identify preventive measures.

Site managers actively seek feedback from the construction team to identify areas for improvement in safety protocols and procedures. They involve workers in safety committees or toolbox talks, encouraging active participation and ownership of safety initiatives.

Addressing safety violations and non-compliance is a fundamental responsibility of site managers to maintain a safe and secure construction site. By taking immediate action, implementing corrective measures, and promoting a culture of safety, site managers demonstrate their unwavering commitment to the well-being of the construction team. Addressing safety violations is not just a regulatory requirement but a testament to the strong leadership and dedication of site managers in creating a safe working environment where every individual can perform their tasks with confidence and peace of mind.

Providing Continuous Safety Training for Workers

In the domain of construction site management, ensuring the safety and well-being of the workforce is a top priority. Aspiring and current site managers recognize that continuous safety training is a cornerstone of maintaining a strong safety culture and fostering a proactive approach towards risk prevention. Through ongoing education and reinforcement of safety protocols, site managers empower workers with the knowledge and skills to navigate potential hazards confidently.

Key Elements of Continuous Safety Training

Continuous safety training encompasses various elements designed to cater to the unique challenges and requirements of the construction site. Some key aspects of this training include:

Regular Refresher Courses: Site managers organize regular refresher courses for workers to reinforce fundamental safety principles. These sessions cover essential topics such as proper use of personal protective equipment (PPE), hazard identification, emergency response protocols, and safe work practices for specific tasks.

Task-Specific Training: Construction sites often involve diverse tasks and specialized activities. Site managers provide task-specific safety training tailored to the unique risks associated with each job. Whether it's operating heavy machinery, working at heights, or handling hazardous materials, workers receive targeted instruction and guidance to execute their tasks safely.

Updated Industry Best Practices: Safety standards and practices in the construction industry continually evolve. Site managers ensure that workers stay abreast of the latest industry best practices through continuous learning opportunities. This may involve attending seminars, workshops, or webinars conducted by safety experts and industry professionals.

Simulation and Hands-On Training: In addition to theoretical instruction, site managers incorporate simulation and hands-on training to offer a practical understanding of safety procedures. Through realistic scenarios and exercises, workers gain valuable experience in managing emergency situations and making informed decisions in a safe environment.

Toolbox Talks and Safety Meetings: Site managers conduct regular toolbox talks and safety meetings to facilitate open communication and knowledge sharing among the construction team. These discussions address specific safety concerns, share lessons learned from past incidents, and encourage workers to contribute their insights to improve safety practices.

Benefits of Continuous Safety Training

Continuous safety training yields several benefits that contribute to the overall success of a construction project and the well-being of the workforce:

Enhanced Safety Awareness: Workers develop a heightened sense of safety awareness, enabling them to proactively identify and address potential hazards. This leads to a more vigilant and cautious approach to work, reducing the likelihood of accidents and injuries.

Increased Confidence: Properly trained workers feel more confident in their ability to perform tasks safely. This confidence fosters a positive work environment where workers are empowered to take ownership of their safety and that of their colleagues.

Reduced Downtime and Losses: Through continuous safety training, workers become more adept at preventing accidents and minimizing disruptions. This, in turn, results in reduced downtime, increased productivity, and lower financial losses associated with accidents and delays.

Improved Team Morale: A commitment to continuous safety training demonstrates that site managers prioritize the well-being of their workforce. This approach fosters a culture of care and respect, boosting team morale and promoting a cohesive and motivated construction team.

Compliance and Legal Requirements: Continuous safety training ensures that the construction site remains compliant with regulatory and legal requirements. This not only avoids potential penalties but also reinforces the site's reputation as a responsible and ethical operation.

Providing continuous safety training for workers is an integral part of effective site management in the construction industry. By investing in ongoing education, site managers equip their workforce with the knowledge and skills needed to work safely and proactively address potential risks. Continuous safety training enhances safety awareness, boosts confidence, reduces accidents, and strengthens the overall safety culture on the construction site. It is a testament to the site manager's commitment to prioritizing the well-being of the workforce and ensuring the successful and secure execution of the construction project.

CHAPTER 5: QUALITY CONTROL AND INSPECTION

I n this chapter, we consider into the critical aspect of quality control and inspection in construction site management. Aspiring and current site managers recognize the paramount importance of maintaining high-quality standards throughout the construction process. By implementing rigorous quality control measures and conducting comprehensive inspections, site managers ensure that construction work meets the specified requirements and adheres to industry best practices. This chapter explores the key elements of quality control, setting quality standards, performing inspections, and implementing corrective actions to deliver exceptional outcomes in construction projects.

Setting Quality Standards
for Construction Work

Ensuring high-quality construction work is a top priority for aspiring and current site managers. To achieve this, site managers play a crucial role in setting clear and measurable quality standards from the project's outset. These standards act as a roadmap for the construction team, guiding them towards delivering work that meets or exceeds the expectations of clients, stakeholders, and regulatory bodies.

Understanding Project Requirements

Site managers begin by thoroughly understanding the project requirements, specifications, and design documents. They collaborate closely with project managers, architects, engineers, and other stakeholders to gain a comprehensive understanding of the project's objectives and the desired quality of the final outcome.

Defining Acceptance Criteria

Based on the project requirements, site managers work with the team to define acceptance criteria for different aspects of the construction work. Acceptance criteria specify the minimum acceptable quality level for various components of the project, such as structural integrity, finishing, safety standards, and functionality.

Aligning with Industry Standards

Site managers ensure that the quality standards set for the construction work align with industry best practices and relevant building codes. Adherence to established industry standards is crucial to ensuring that the construction work meets safety, durability, and performance benchmarks.

Implementing Quality Assurance Procedures

To uphold quality standards, site managers establish comprehensive quality assurance procedures. These procedures include regular inspections, testing, and verification of materials and workmanship. Quality assurance teams conduct thorough checks to identify and address any deviations from the established quality standards.

Continuous Monitoring and Feedback

Site managers emphasize continuous monitoring of the

construction work at various stages. They provide timely feedback to the construction team, addressing any shortcomings and acknowledging areas of excellence. Regular feedback loops help maintain a consistent focus on achieving and surpassing quality expectations.

Promoting a Culture of Excellence

Beyond technical aspects, site managers foster a culture of excellence in the construction team. They emphasize the importance of taking pride in one's work and the significance of delivering high-quality outcomes. This culture of excellence motivates workers to take ownership of the quality of their work and strive for continuous improvement.

Addressing Non-Conformances

Despite proactive measures, non-conformances may arise during construction. Site managers address these issues promptly and implement corrective actions to rectify deviations from quality standards. They engage with the team to identify the root causes of non-conformances and work collaboratively to prevent their recurrence.

Client Satisfaction and Feedback

Site managers value client satisfaction and seek feedback from clients throughout the construction process. Client feedback provides valuable insights into meeting or surpassing their expectations. It helps site managers gauge the success of the quality standards implemented and make necessary adjustments if required.

Setting quality standards for construction work is a fundamental responsibility of site managers. By establishing clear and measurable expectations, aligning with industry standards, and promoting a culture of excellence, site managers ensure that construction projects are executed to the highest quality possible. Emphasizing continuous monitoring, feedback, and addressing non-conformances allows site managers to deliver exceptional results that reflect their commitment to excellence and client satisfaction. Quality control remains an integral part of effective site management, contributing to the success and reputation of construction projects and the entire construction team.

Defining Acceptance Criteria for Workmanship and Materials

In construction site management, defining clear acceptance criteria for workmanship and materials is vital to ensure that the quality of construction meets the specified standards and requirements. Aspiring and current site managers play a pivotal role in setting these criteria, which act as benchmarks for assessing the performance and quality of both the workforce and the materials used in the construction process.

Workmanship Acceptance Criteria

Workmanship acceptance criteria encompass the standards and expectations for the craftsmanship and execution of construction tasks. Site managers collaborate with architects, engineers, and project managers to define specific parameters for workmanship, including:

Precision and Accuracy: Ensuring that construction work is precise, accurate, and in strict compliance with architectural and engineering drawings.

Finishing Quality: Defining the expected quality of surface finishes, whether it pertains to walls, flooring, ceilings, or other elements of the construction.

Joint and Seam Integrity: Setting standards for the integrity of joints and seams, especially in critical areas such as structural connections and waterproofing elements.

Attention to Detail: Emphasizing the importance of attention to detail in all construction tasks, ensuring that no aspect is overlooked or inadequately executed.

Safety Compliance: Ensuring that workmanship adheres to all safety protocols and regulations, promoting a safe working environment for the construction team.

Materials Acceptance Criteria

Materials acceptance criteria define the quality, durability, and performance standards for construction materials to be used in the project. Site managers collaborate with material suppliers and experts to establish these criteria, which may include:

Material Specifications: Outlining the specific requirements and standards for various materials, such as concrete, steel, timber, and finishes.

Quality Testing: Implementing quality testing procedures to verify the properties and characteristics of materials before their use in construction.

Environmental Sustainability: Ensuring that materials comply with environmental sustainability standards, promoting eco-friendly construction practices.

Longevity and Durability: Setting criteria for the expected lifespan and durability of materials to ensure the longevity of the constructed assets.

Compliance with Regulations: Ensuring that all materials meet relevant building codes, safety regulations, and industry standards.

Inspection and Verification

Site managers conduct regular inspections and verification processes to assess whether the workmanship and materials meet the defined acceptance criteria. Inspections may involve visual assessments, non-destructive testing, and material sampling. Any deviations from the criteria are promptly addressed, and corrective actions are implemented to maintain quality standards.

Documenting Acceptance

Site managers document the acceptance criteria for workmanship and materials in the project's quality management plan. These documented criteria serve as a reference for the construction team and help in evaluating the overall progress and quality of the project.

By defining clear acceptance criteria for workmanship and materials, site managers provide a standardized framework to assess and uphold the quality of construction work. This commitment to maintaining high standards ensures that the construction project not only meets the desired quality but also establishes the reputation of the construction team and site manager as a reliable and proficient entity in the construction industry.

Establishing Quality Control Processes and Procedures

Quality control is a fundamental aspect of effective construction site management, ensuring that construction work meets the specified standards and delivers exceptional outcomes. Aspiring and current site managers play a critical role in establishing comprehensive quality control processes and procedures to maintain consistent and reliable construction quality. Here, we outline the key steps involved in setting up robust quality control measures:

Developing a Quality Management Plan

Site managers begin by developing a comprehensive quality management plan that outlines the approach to achieving and maintaining high-quality standards throughout the project. The plan includes a detailed description of the quality control processes, the roles and responsibilities of the quality control team, and the procedures for inspecting, testing, and verifying workmanship and materials.

Defining Quality Control Checkpoints

Site managers collaborate with architects, engineers, project managers, and other stakeholders to identify critical checkpoints during various stages of construction. These checkpoints serve as milestones for evaluating the quality of work at different project phases. They include assessments of structural integrity, finishes, safety compliance, and adherence to design specifications.

Implementing Inspection and Testing Protocols

Site managers set up systematic inspection and testing protocols to assess the quality of construction work. These protocols encompass regular inspections by trained personnel and, when necessary, testing of materials to ensure they meet the specified requirements. Non-destructive testing methods are employed to verify the integrity of critical structural components without compromising the construction's stability.

Documenting Quality Control Findings

Site managers meticulously document all quality control findings, including inspection reports, test results, and non-conformance reports. This documentation provides a record of the construction quality and serves as evidence of compliance with industry standards and regulations.

Implementing Corrective and Preventive Actions

When quality control identifies deviations or non-conformities, site managers take prompt corrective actions to rectify the issues. They collaborate with the construction team to address the root causes and implement measures to prevent similar occurrences in the future. Preventive actions are also put in place to mitigate potential quality risks.

Conducting Quality Audits

Regular quality audits are conducted by independent parties or internal quality control teams to assess the overall effectiveness of the quality control processes and procedures. Audits help identify areas for improvement and ensure that quality standards are consistently upheld.

Continuous Improvement
Site managers promote a culture of continuous improvement in quality control. They encourage feedback from the construction team, stakeholders, and clients to identify opportunities for enhancing quality control processes and procedures.

Training and Professional Development
To ensure the competence of the quality control team, site managers prioritize training and professional development. Regular training programs keep the team abreast of new industry standards, best practices, and innovative quality control techniques.

By establishing robust quality control processes and procedures, site managers demonstrate their commitment to delivering construction work of the highest caliber. The systematic approach to quality control instills confidence in clients, stakeholders, and the construction team, making it a key element in the success of any construction project.

Incorporating Quality into the Project Management Framework
Integrating quality into the project management framework is essential for ensuring that construction projects meet the highest standards and deliver successful outcomes. Aspiring and current site managers play a crucial role in embedding quality considerations throughout the project lifecycle. Here's how quality can be seamlessly integrated into the project management framework:

1. Establishing Quality Objectives
At the project's inception, site managers collaborate with project stakeholders to define clear quality objectives. These objectives

align with the project's overall goals and specify the desired level of quality for different aspects of the construction work. Quality objectives may include meeting regulatory requirements, achieving certain performance standards, or exceeding client expectations.

2. Identifying Critical Quality Factors

Site managers work with architects, engineers, and project teams to identify critical quality factors that significantly impact the project's success. These factors could encompass structural integrity, safety compliance, adherence to design specifications, workmanship, and materials. By identifying these factors, site managers prioritize their focus on areas with the highest quality impact.

3. Integrating Quality Activities into Project Planning

During the project planning phase, site managers incorporate quality-related activities into the project schedule and budget. They allocate resources for quality control inspections, testing, and documentation. Quality checkpoints are integrated into the project timeline to assess progress and compliance with quality standards at specific intervals.

4. Engaging Stakeholders in Quality Assurance

Site managers engage stakeholders, including clients, architects, engineers, contractors, and regulatory authorities, in the quality assurance process. Regular communication ensures that all parties understand the quality requirements and contribute to maintaining high standards throughout the project.

5. Risk Management for Quality

Quality considerations are integrated into the project's risk management framework. Site managers identify potential quality risks and develop risk mitigation strategies to prevent or address any quality-related challenges that may arise during construction.

6. Collaborative Decision-Making for Quality

Site managers foster a collaborative decision-making approach that involves the entire project team. Decisions related to materials, construction methods, and design modifications are made collectively with quality as a central consideration.

7. Documenting Quality Control and Assurance

Quality-related documentation is an integral part of the project management framework. Site managers ensure that all quality control activities, inspection reports, testing results, and non-conformance reports are accurately recorded and communicated to relevant stakeholders.

8. Continuous Monitoring and Improvement

Throughout the project's execution, site managers continuously monitor and assess the quality of construction work. They use performance indicators and metrics to measure the project's adherence to quality objectives. Any deviations or areas for improvement are identified and addressed promptly.

9. Closing the Loop

After project completion, site managers conduct a comprehensive review to evaluate the overall quality of the construction work. Lessons learned are documented and incorporated into future projects to continuously improve the quality management process.

By seamlessly incorporating quality into the project management framework, site managers demonstrate their commitment to delivering construction projects of the highest quality. This approach ensures that quality considerations are not treated as an afterthought but rather integrated into every phase of the project, resulting in successful and sustainable construction outcomes.

Performing Inspections and Quality Assurance Checks

Conducting thorough inspections and quality assurance checks is a fundamental responsibility of site managers to ensure that construction projects adhere to the highest standards of quality. These activities are essential in identifying potential issues, verifying compliance with specifications, and maintaining the overall integrity of the project. Here's how site managers perform inspections and quality assurance checks:

1. Establishing Inspection Protocols

Site managers create detailed inspection protocols that outline the specific criteria and requirements for each aspect of the construction work. These protocols are based on industry standards, regulatory guidelines, and project-specific quality objectives.

2. Regular On-Site Inspections

Site managers conduct regular on-site inspections at various stages of the construction process. They meticulously examine the workmanship, materials, and adherence to design specifications. Through visual assessments and measurements, site managers verify that the construction meets the established quality standards.

3. Documenting Inspection Findings:

During inspections, site managers document their findings thoroughly. They create detailed reports that include observations, deviations from standards, and any non-conformances identified. These reports serve as a critical reference for corrective actions and quality improvement.

4. Implementing Corrective Actions

When issues or non-conformances are identified, site managers take prompt corrective actions. They work closely with the project team to address the root causes of problems and implement appropriate solutions to bring the work back to compliance.

5. Non-Destructive Testing (NDT)

Site managers may engage specialized NDT techniques to assess the integrity of critical structures or materials without causing damage. NDT methods include ultrasonic testing, radiographic testing, and magnetic particle inspection, among others.

6. Conducting Quality Assurance Checks
In addition to inspections, site managers perform quality assurance checks on materials and equipment used in the construction. They ensure that all materials meet the required standards and are sourced from approved suppliers.

7. Collaboration with Quality Control Teams
Site managers work closely with dedicated quality control teams, if present, to align inspection efforts and verify that quality control processes are integrated into the project management framework.

8. Compliance with Safety Standards
During inspections and quality assurance checks, site managers also verify compliance with safety standards and regulations. This helps ensure a safe working environment for construction personnel and minimizes potential safety risks.

9. Continuous Improvement and Lessons Learned
Site managers use the insights gained from inspections and quality assurance checks to continuously improve construction processes and prevent future issues. Lessons learned from previous projects are integrated into current practices to enhance overall project quality.

Performing inspections and quality assurance checks is an ongoing commitment by site managers to maintain the highest level of quality throughout the construction project. By diligently examining and verifying each aspect of the work, site managers contribute significantly to the successful delivery of safe and durable construction projects.

Conducting Routine and Detailed Inspections

Routine and detailed inspections are essential practices that site managers undertake to ensure the quality, safety, and compliance of construction projects. These inspections play a crucial role in identifying potential issues early on, verifying adherence to specifications, and maintaining overall project integrity. By conducting routine inspections at predefined intervals and carrying out detailed assessments of critical components, site managers can effectively monitor progress and take timely corrective actions when needed.

During routine inspections, site managers systematically review various aspects of the construction project, including workmanship, materials usage, safety measures, and adherence to design specifications. Regular assessments enable them to monitor the ongoing work and detect any deviations from the project plan or quality standards. These inspections are scheduled at specific intervals, allowing site managers to keep a close eye on progress and identify emerging challenges.

Detailed inspections, on the other hand, involve in-depth examinations of specific areas or critical components of the project. Site managers conduct these inspections when they need a more comprehensive assessment of a particular aspect of the work. Whether it's assessing complex systems, intricate structural components, or specialized installations, detailed inspections help ensure precision and compliance with stringent requirements.

Documentation and reporting are integral to the inspection process. After conducting inspections, site managers meticulously document their findings, observations, and any non-conformances. These reports serve as valuable records of the project's progress and form the basis for communicating critical information to stakeholders, project teams, and regulatory authorities. They also facilitate the development of action plans and the implementation of necessary corrective measures.

Collaboration and effective communication are paramount during inspections. Site managers work closely with project teams, subcontractors, and quality control personnel to convey inspection findings and discuss potential issues. Collaborative efforts foster a proactive approach to address challenges and maintain project quality.

Moreover, routine and detailed inspections are closely integrated with quality control efforts. Site managers collaborate with dedicated quality control teams to align inspection activities with overall quality assurance processes. This integration ensures that construction work consistently meets the specified quality standards.

Continuous improvement is another key aspect of inspections. Site managers leverage insights gained from routine and detailed inspections to enhance construction practices continuously. Lessons learned from previous inspections are applied to streamline operations and minimize potential risks in future projects.

Ensuring Compliance with Design Specifications

One of the fundamental responsibilities of a site manager is to ensure strict compliance with design specifications throughout the construction process. Design specifications serve as the blueprint for the project, outlining every detail required to achieve the desired outcome. It is the site manager's duty to meticulously follow these specifications to ensure the project meets the intended design and performance criteria.

To ensure compliance with design specifications, site managers undertake several crucial tasks. First and foremost, they thoroughly review the design documents, including architectural plans, engineering drawings, and technical specifications. This detailed review allows site managers to gain a comprehensive understanding of the project's requirements, materials, and construction methods.

Next, site managers collaborate closely with the design teams and project stakeholders to clarify any ambiguities or address potential conflicts in the specifications. Effective communication between all parties involved is vital in resolving design-related issues early on and preventing costly modifications during construction.

As construction progresses, site managers continuously monitor the work to verify that it aligns with the approved design specifications. They conduct regular inspections to assess the quality of workmanship, materials used, and overall adherence to the design intent. Any deviations from the specifications are promptly addressed, and corrective actions are taken to bring the project back on track.

In addition to inspections, site managers implement rigorous quality control measures to ensure that each construction component meets the specified standards. This includes testing and verification of critical elements, such as structural integrity, electrical systems, plumbing, and HVAC installations.

Site managers also play a vital role in coordinating with subcontractors and ensuring that they are well-informed about the design requirements relevant to their scope of work. Clear communication and coordination with subcontractors are essential to avoid misunderstandings and ensure uniformity in meeting design specifications.

Technology plays an increasingly important role in ensuring compliance with design specifications. Site managers leverage advanced construction management software and BIM tools to precisely track the progress of work, identify potential clashes or deviations, and facilitate real-time collaboration with all stakeholders.

Lastly, documentation is a crucial aspect of ensuring compliance with design specifications. Site managers maintain detailed records of all design-related communications, inspection reports, and quality control measures. These records not only serve as a reference for the current project but also provide valuable insights for future endeavors and lessons learned.

Documenting Findings and Implementing Corrective Actions
Documenting findings and implementing corrective actions are critical components of a site manager's role in maintaining quality and addressing any non-conformances identified during inspections and quality control processes.

When site managers conduct routine and detailed inspections, they meticulously document their observations, noting any discrepancies between the actual work and the specified design standards. These documented findings serve as a comprehensive record of the project's progress, enabling site managers to track the quality of work over time and identify recurring issues or trends.

Accurate and detailed documentation allows site managers to communicate effectively with the project team, subcontractors, and other stakeholders. By providing clear and concise reports, site managers facilitate transparent communication about the project's status, potential challenges, and the necessary corrective actions.

When non-conformances or deviations from design specifications are identified, site managers must take prompt and decisive action. This involves implementing corrective measures to rectify the issues and bring the project back into compliance. Depending on the nature and severity of the non-conformance, corrective actions may range from minor adjustments to significant modifications in the construction process.

Site managers work closely with the project team and subcontractors to develop action plans for addressing non-conformances. These action plans outline specific steps to be taken, responsibilities assigned to individuals, and timelines for completion. Regular follow-up and monitoring ensure that corrective actions are carried out efficiently and effectively.

In addition to addressing immediate non-conformances, site managers also assess the root causes of these issues. By understanding the underlying reasons for deviations from design specifications, they can implement preventive measures to avoid similar problems in the future. This proactive approach to quality management helps to continuously improve construction processes and ensure consistent adherence to design standards.

Site managers use their documented findings to provide valuable feedback to the project team and subcontractors. Constructive feedback allows for a better understanding of expectations and promotes a culture of continuous improvement and learning among all stakeholders.

Effective documentation and corrective action implementation are also essential for complying with regulatory requirements and contractual obligations. By maintaining detailed records of inspections and quality control measures, site managers can demonstrate due diligence and compliance in case of audits or legal issues.

Managing Non-Conformances and Corrective Actions
Managing non-conformances and corrective actions is a fundamental aspect of a site manager's responsibility in ensuring the quality and compliance of construction projects. Non-conformances refer to instances where the work deviates from design specifications or established quality standards. When non-conformances are identified during inspections or quality control processes, site managers must initiate

a systematic approach to address and resolve these issues promptly and effectively.

The first step in managing non-conformances is to document and categorize them based on their nature, severity, and impact on the project. This systematic documentation helps site managers prioritize corrective actions and allocate resources accordingly. It also aids in identifying potential patterns or recurring issues that may require more comprehensive solutions.

Upon identifying non-conformances, site managers work collaboratively with the project team, subcontractors, and quality control personnel to analyze the root causes. Understanding the underlying reasons behind the non-conformances is crucial in developing appropriate corrective actions and preventing similar issues in the future. This analysis may involve conducting further investigations, utilizing testing methods, or seeking expert input, depending on the complexity of the non-conformance.

Site managers then develop comprehensive action plans that outline the specific steps to be taken to address each non-conformance. These action plans include clear objectives, responsibilities assigned to individuals or teams, and realistic timelines for completion. In complex situations, site managers may establish cross-functional teams to coordinate efforts and ensure a holistic approach to resolving the non-conformances.

Regular monitoring and tracking of the corrective actions are essential to verify their effectiveness and progress. Site managers conduct follow-up inspections and reviews to assess the implementation of the action plans and confirm that the non-conformances have been adequately addressed. Any deviation from the action plan is promptly addressed, and adjustments are made as needed.

Additionally, effective communication throughout the process is vital to keep all stakeholders informed about the status of the non-conformances and the progress of the corrective actions. Open and transparent communication fosters a collaborative environment, encourages accountability, and promotes a shared commitment to achieving quality outcomes.

Site managers also recognize the importance of continuous improvement. They use insights gained from managing non-conformances to update and refine construction processes, quality control measures, and training programs. By incorporating lessons learned, site managers contribute to the ongoing enhancement of construction practices and the prevention of future non-conformances.

Managing non-conformances and corrective actions is an integral part of a site manager's responsibility to ensure the successful delivery of high-quality construction projects. By employing systematic documentation, root cause analysis, comprehensive action plans, and regular monitoring, site managers can address non-conformances efficiently and improve construction practices for enhanced project outcomes and client satisfaction.

Addressing Quality Deficiencies and Non-Conformities
Addressing quality deficiencies and non-conformities is a fundamental responsibility of site managers to ensure the successful execution of construction projects with the highest standards of quality and compliance. Quality deficiencies refer to shortcomings in workmanship, materials, or processes that fall short of specified requirements, while non-conformities are instances where the work deviates from design specifications or established quality standards.

When quality deficiencies or non-conformities are identified, site managers must take swift and effective action to rectify the issues and ensure that the project proceeds according to the approved plans. This process requires a systematic approach and a focus on problem-solving to ensure that the project meets the expected level of quality and safety.

The first step in addressing quality deficiencies and non-conformities is a thorough assessment. Site managers conduct detailed inspections and reviews of documentation to identify all areas where the project may not meet the required standards. This assessment may involve collaboration with quality control personnel and consultation with the project team to gather insights and observations.

Once the issues have been identified, site managers prioritize them based on their severity and potential impact on the project. Critical or safety-related non-conformities are given immediate attention to ensure the safety of workers and the integrity of the structure.

To effectively address the identified deficiencies and non-conformities, site managers conduct a root cause analysis. Understanding the underlying reasons for the deviations from quality standards or design specifications is crucial to implement appropriate corrective actions and prevent similar issues in the future.

Site managers work closely with the project team, subcontractors, and quality control personnel to develop comprehensive corrective action plans. These plans outline specific steps to be taken, responsibilities assigned to individuals or teams, and realistic timelines for completion.

The implementation of corrective actions is closely monitored by site managers. They ensure that the necessary adjustments are made promptly and effectively to resolve the identified issues. Additionally, site managers may introduce quality assurance measures to prevent similar problems from occurring in the future.

Throughout the process, effective communication with all

stakeholders is vital. Site managers maintain open and transparent communication with clients, subcontractors, and regulatory authorities to keep them informed of the actions being taken to address the quality deficiencies and non-conformities.

Addressing quality deficiencies and non-conformities is not just about solving immediate problems but also about continuous improvement. Site managers use insights gained from addressing these issues to refine construction processes, update quality control measures, and implement best practices to enhance project outcomes.

Implementing Corrective and Preventive Actions

Implementing corrective and preventive actions is a critical aspect of effective site management to ensure the successful completion of construction projects with the highest level of quality and compliance. Corrective actions focus on addressing specific issues or non-conformities that have already occurred during project execution, while preventive actions aim to identify and mitigate potential risks before they lead to problems. Both types of actions play a pivotal role in maintaining a smooth and efficient construction process.

When issues arise during the construction phase, site managers must promptly initiate corrective actions. This involves a systematic approach to identify the root cause of the problem and develop tailored solutions to rectify it. Site managers collaborate with the project team, subcontractors, and relevant stakeholders to ensure that the corrective actions align with project goals and meet quality standards. Assigning clear responsibilities and ensuring effective communication are essential in implementing corrective actions efficiently.

A comprehensive root cause analysis is essential to understand why the issue occurred and to prevent its recurrence. Site managers use this analysis to develop detailed corrective action plans that outline the specific steps to be taken, the responsible parties, and the expected timeline for implementation. By adhering to these action plans, site managers can effectively address the issues and maintain project momentum.

In addition to addressing existing issues, site managers must be proactive in implementing preventive actions. This involves conducting thorough risk assessments to identify potential hazards and risks that may affect the project. By anticipating and understanding potential risks, site managers can develop strategies to mitigate them before they materialize.

Preventive actions require ongoing monitoring and vigilance throughout the project's duration. Site managers continually observe project progress and activities to identify any emerging risks. This enables them to adapt and refine the preventive actions as needed based on changing circumstances.

By effectively implementing both corrective and preventive actions, site managers demonstrate their commitment to delivering high-quality projects while ensuring the safety of workers and compliance with regulations. Proactive risk management and swift resolution of issues are fundamental in maintaining project success and earning the trust and satisfaction of clients and stakeholders.

Tracking Quality Improvement Initiatives

Tracking quality improvement initiatives is an integral part of effective site management, as it enables site managers to monitor the progress and effectiveness of measures taken to enhance the overall quality of construction projects. Quality improvement initiatives are designed to continuously identify areas for enhancement, implement changes, and evaluate the results. By tracking these initiatives, site managers can ensure that project quality meets or exceeds the defined standards and that any necessary adjustments are promptly made.

One of the primary steps in tracking quality improvement

initiatives is establishing clear and measurable objectives. Site managers define specific goals and targets for improvement in various aspects of the project, such as workmanship, materials, safety, and adherence to design specifications. These objectives should be realistic, achievable, and aligned with the overall project goals.

To track progress, site managers collect relevant data and information related to the quality improvement initiatives. This data may include inspection reports, quality control measurements, feedback from stakeholders, and records of any corrective actions taken. By systematically gathering and organizing this data, site managers can gain valuable insights into the effectiveness of the initiatives and identify areas that require further attention.

Site managers should also establish key performance indicators (KPIs) to measure the success of quality improvement initiatives. KPIs provide quantifiable metrics that allow site managers to gauge progress and compare results against predefined benchmarks. Some common KPIs for tracking quality improvement include defect rates, customer satisfaction ratings, safety incident rates, and adherence to project schedules.

Regular monitoring and reporting are essential aspects of tracking quality improvement initiatives. Site managers conduct periodic reviews to assess whether the established objectives are being met and to identify any deviations or challenges. These reviews also provide an opportunity to recognize successful initiatives and share best practices with the project team.

In addition to internal tracking, site managers may collaborate with external quality assurance entities or conduct independent third-party audits to validate the project's quality and compliance with industry standards. External evaluations provide an impartial assessment of the project's performance and offer valuable feedback for continuous improvement.

As site managers track quality improvement initiatives, they must be proactive in implementing corrective actions when deviations or deficiencies are identified. These actions may involve revising processes, providing additional training, or making changes to materials or equipment. The aim is to continually refine project practices and enhance the overall quality of construction deliverables.

CHAPTER 6:
MANAGING
SUBCONTRACTORS
AND TRADES

Managing subcontractors and trades is a critical aspect of site management, as construction projects often involve multiple specialized contractors and tradespeople. Effectively overseeing these external entities ensures smooth coordination, adherence to project schedules, and the successful completion of tasks within budget and quality standards. Site managers play a pivotal role in fostering positive relationships with subcontractors and trades to create a collaborative and efficient working environment. Through clear communication, contract administration, and proactive issue resolution, site managers optimize subcontractor performance and contribute to the overall success of the project.

Selecting and Engaging
Reliable Subcontractors

Selecting and engaging reliable subcontractors is a crucial responsibility of site managers in construction projects.

Subcontractors play a vital role in executing specialized tasks and contributing to the overall project's success. Therefore, site managers must follow a rigorous selection process to ensure that the subcontractors they choose are competent, reliable, and aligned with the project's requirements.

The first step in selecting subcontractors is to identify the specific tasks or trades that require external expertise. Site managers assess the project's scope and determine which aspects can be best handled by specialized subcontractors. They then create a list of potential subcontractors based on their reputation, experience, and track record in similar projects.

Once the list is compiled, site managers conduct a thorough evaluation of each subcontractor. This evaluation involves reviewing their past projects, checking references, and verifying licenses and certifications. Site managers also assess the subcontractors' financial stability to ensure they can fulfill their contractual obligations.

Communication and collaboration are key factors in engaging reliable subcontractors. Site managers discuss the project's expectations, timelines, and deliverables with potential subcontractors. They ensure that the subcontractors fully understand the scope of work and are committed to meeting the project's requirements.

Negotiating clear and fair contract agreements is essential to avoid misunderstandings and disputes. Site managers define the scope of work, payment terms, timelines, and performance expectations in the contract. They also include provisions for any potential changes or unforeseen circumstances that may arise during the project.

Throughout the project, site managers maintain open lines of communication with subcontractors. Regular meetings and progress updates help ensure that the subcontractors are on track with their tasks and that any issues or challenges are promptly addressed.

Site managers also foster a collaborative and positive working relationship with subcontractors. They recognize and acknowledge the subcontractors' contributions to the project's success, which can boost morale and motivate them to perform at their best.

Evaluating Subcontractor Qualifications and Performance

Evaluating subcontractor qualifications and performance is a fundamental responsibility of site managers to ensure the successful execution of construction projects. Site managers must thoroughly assess potential subcontractors' qualifications before engaging them and continuously monitor their performance throughout the project's duration.

To evaluate subcontractor qualifications, site managers review their expertise, experience, and credentials in the specific trade or task required for the project. This assessment involves verifying licenses, certifications, and affiliations with relevant industry organizations. Site managers also examine the subcontractor's past project history to gauge their capability to deliver high-quality work.

In addition to qualifications, site managers assess subcontractor capacity and resources. They determine whether the subcontractor has sufficient manpower, equipment, and materials to complete the assigned tasks within the project's timeline. Adequate resources are essential to avoid delays and ensure project milestones are met.

Another critical aspect of evaluating subcontractors is checking their financial stability. Site managers verify the subcontractor's financial standing to ensure they can fulfill their contractual obligations, including paying their workforce and procuring necessary materials.

Once engaged in the project, site managers closely monitor

subcontractor performance. Regular site visits, progress reports, and quality inspections help assess how well the subcontractor is meeting project requirements. Any deviations from the agreed-upon scope of work, schedule delays, or quality issues are promptly addressed to prevent further complications.

Communication is a vital tool in evaluating subcontractor performance. Site managers maintain open and transparent communication with subcontractors to discuss any challenges, provide feedback, and offer support when needed. This collaborative approach fosters a positive working relationship and allows for timely resolution of issues.

Site managers may also implement performance metrics and key performance indicators (KPIs) to objectively measure subcontractor performance. KPIs could include adherence to schedule, quality of work, safety record, and adherence to contract terms.

In situations where a subcontractor's performance falls below expectations, site managers take corrective actions to address the issues. This may involve providing additional support, retraining, or, in extreme cases, replacing the subcontractor with a more qualified and reliable one.

Evaluating subcontractor qualifications and performance is essential for site managers to ensure the successful execution of construction projects. By conducting thorough assessments before engagement and implementing continuous monitoring and communication during the project, site managers can maintain high standards of workmanship, adhere to project timelines, and deliver successful outcomes. A proactive and collaborative approach in evaluating subcontractors contributes significantly to the overall project's success and the satisfaction of all stakeholders involved.

Negotiating Clear Contract Agreements and Terms

Negotiating clear contract agreements and terms is a crucial aspect of site management, as it establishes the foundation for successful collaboration between the construction project's parties. Site managers play a central role in ensuring that all parties involved clearly understand their roles, responsibilities, and obligations throughout the project's lifecycle.

During contract negotiations, site managers work closely with subcontractors, suppliers, and other stakeholders to define the scope of work, deliverables, and timelines. They carefully outline the specific tasks and services to be provided by each party and set measurable performance metrics to gauge progress and success.

To create clear and effective contract agreements, site managers must pay meticulous attention to details. They ensure that all terms and conditions are accurately documented, leaving no room for ambiguity or misunderstandings. Key contract elements include payment terms, warranties, dispute resolution procedures, and any specific provisions related to project changes or unforeseen circumstances.

Site managers also address risk management in the contract negotiations, outlining procedures for identifying, mitigating, and allocating risks among the parties involved. This proactive approach helps prevent potential disputes and ensures that all parties are aware of the potential risks they may face during the project.

In addition to outlining responsibilities and obligations, site managers negotiate fair and equitable compensation for subcontractors and suppliers. They strive to strike a balance between meeting budget constraints and providing reasonable compensation that aligns with the level of expertise and quality expected from each subcontractor.

Throughout the negotiation process, site managers maintain open communication with all parties involved, encouraging transparency and trust. They actively listen to the concerns and perspectives of subcontractors and suppliers, addressing any issues promptly to create a positive working environment.

Once the contract is finalized and agreed upon, site managers ensure that all parties adhere to the terms and conditions outlined. They continuously monitor compliance and promptly address any deviations or potential breaches. Site managers also keep detailed records of all communications and actions related to the contract to protect all parties' interests.

Cultivating Positive Relationships with Subcontractors

Cultivating positive relationships with subcontractors is a key priority for site managers as it fosters a collaborative and productive work environment. Building strong partnerships with subcontractors is essential to ensure the successful execution of construction projects and to achieve the project's objectives efficiently.

One of the first steps in cultivating positive relationships is selecting reliable and competent subcontractors. Site managers carefully evaluate and choose subcontractors based on their qualifications, experience, and track record. Working with reputable and skilled subcontractors sets a solid foundation for a positive working relationship.

Effective communication is at the heart of building positive relationships. Site managers maintain open and transparent lines of communication with subcontractors. They actively listen to their concerns, feedback, and suggestions, ensuring that the subcontractors feel valued and heard. Regular meetings and clear communication channels help address any issues promptly and avoid misunderstandings.

Mutual respect is a fundamental aspect of fostering positive relationships. Site managers treat subcontractors with respect, recognizing their expertise and contribution to the project. In turn, subcontractors are more likely to be committed to the project's success and perform at their best when they feel respected and appreciated.

Providing clear expectations and guidelines is essential for a smooth working relationship. Site managers communicate project objectives, timelines, and quality standards effectively to subcontractors. Clarity in expectations helps avoid confusion and minimizes the likelihood of conflicts arising from misunderstandings.

Site managers also support subcontractors in their work. They offer assistance and resources when needed, ensuring that subcontractors have the necessary tools and information to perform their tasks effectively. This support fosters a sense of teamwork and collaboration among all project participants.

Recognizing and acknowledging subcontractors' achievements and contributions is an effective way to motivate and encourage their performance. Site managers publicly commend subcontractors for their hard work and dedication, fostering a positive and motivating work culture.

In challenging situations or when issues arise, site managers approach conflicts with a problem-solving mindset. Instead of placing blame, they collaborate with subcontractors to identify solutions and work together towards resolution. This approach helps maintain trust and ensures that relationships remain positive, even during difficult times.

Finally, site managers consistently evaluate their own performance as well. They seek feedback from subcontractors on how they can improve their management practices and ensure that they are fulfilling their role in facilitating a successful partnership.

Contract Management and Vendor Coordination

Contract management and vendor coordination are essential components of effective site management in construction projects. Site managers play a vital role in overseeing the contractual agreements with subcontractors, suppliers, and vendors, ensuring that all parties fulfill their obligations and work together harmoniously to achieve project success.

At the outset of a project, site managers meticulously review and finalize the contracts with subcontractors and vendors. They ensure that the contracts are comprehensive, clear, and align with the project's scope and objectives. Site managers pay close attention to details, including payment terms, deliverables, performance metrics, and dispute resolution mechanisms. By having well-structured contracts, site managers establish a solid foundation for effective vendor coordination.

Throughout the project's lifecycle, site managers proactively manage vendor relationships. They maintain open communication with vendors, providing regular updates on project progress and any changes that may impact them. Site managers also address any concerns or challenges promptly, striving to resolve issues in a collaborative manner to minimize disruptions and delays.

Vendor coordination involves managing the flow of materials, equipment, and services required for the project. Site managers work closely with vendors to ensure timely deliveries, adequate inventory levels, and compliance with quality standards. They monitor vendor performance and enforce contractual terms to ensure that vendors meet their commitments.

In cases where adjustments or modifications to contracts are necessary due to unforeseen circumstances or project changes, site managers lead the negotiation and implementation of contract amendments. They aim to strike a balance between protecting the interests of all parties while maintaining the project's schedule and budget.

Effective contract management also involves tracking vendor performance and evaluating their contributions to the project. Site managers assess whether vendors adhere to quality standards, meet deadlines, and provide satisfactory services or products. They provide feedback to vendors and collaborate on areas that require improvement.

In situations where disputes arise, site managers serve as mediators, working to find common ground and resolve conflicts in a fair and equitable manner. They use their negotiation and conflict resolution skills to ensure that all parties find mutually agreeable solutions.

Site managers continuously assess the performance of vendors and subcontractors, identifying opportunities for improvement and efficiency. They seek feedback from project team members and stakeholders to gather insights into vendor performance, enabling them to make informed decisions regarding vendor coordination and future contracts.

Administering Subcontractor Contracts and Deliverables

Administering subcontractor contracts and deliverables is a crucial responsibility of site managers in construction projects. It involves overseeing the execution of contractual agreements, ensuring that subcontractors fulfill their obligations, and verifying that deliverables meet the specified quality and timelines. Effective contract administration contributes to the smooth progress of the project, minimizes potential risks, and helps maintain positive relationships with subcontractors.

Site managers begin by thoroughly reviewing the subcontractor contracts to understand the terms and conditions, scope of work, payment schedules, and performance expectations. They clarify any ambiguities and communicate contract details to subcontractors to ensure mutual understanding and agreement.

Once the contracts are in place, site managers actively monitor and track subcontractor performance. They verify that subcontractors are adhering to project timelines and delivering on agreed-upon milestones. This involves conducting regular inspections, site visits, and progress assessments to assess the quality and completeness of work.

In addition to performance monitoring, site managers handle the administrative aspects of subcontractor contracts. They manage the documentation and record-keeping of all contract-related communications, including change orders, work approvals, and correspondence. Maintaining a well-organized and up-to-date contract file is essential for reference and audit purposes.

Site managers are responsible for processing and approving subcontractor payment requests. They review the invoices submitted by subcontractors, cross-reference them with completed deliverables, and ensure that the charges align with the contracted rates and scope of work. Timely and accurate payment processing is crucial to maintaining positive relationships with subcontractors and ensuring their continued commitment to the project.

When disputes or issues arise between the project team and subcontractors, site managers play a pivotal role in resolving conflicts. They act as mediators, facilitating communication and finding equitable solutions to address disagreements. Effective conflict resolution fosters a cooperative working environment and helps keep the project on track.

In instances where changes to the subcontractor's scope of work are necessary due to project modifications, site managers coordinate with subcontractors to negotiate and implement change orders. They ensure that all changes are documented, agreed upon, and incorporated into the contract to avoid misunderstandings and potential disputes.

Throughout the project's lifecycle, site managers maintain open and transparent communication with subcontractors. They provide regular updates on project progress, communicate any changes that may impact subcontractors, and solicit feedback on ways to enhance collaboration and efficiency.

Facilitating Communication and Collaboration with Subcontractors

Facilitating effective communication and collaboration with subcontractors is a critical responsibility of site managers in the construction industry. As the key liaison between the project's main stakeholders and subcontractors, site managers play a crucial role in ensuring that the project progresses smoothly and meets its objectives.

One of the primary tasks in facilitating communication is establishing clear channels for information exchange. Site managers create an open and transparent environment where subcontractors feel comfortable sharing their insights and raising concerns. Regular meetings and discussions are scheduled to keep all parties informed about project updates, timelines, and potential changes.

To foster collaboration, site managers encourage subcontractors to work together as a cohesive team. They promote a sense of shared purpose, emphasizing the collective responsibility to deliver the project successfully. By instilling a collaborative mindset, site managers enhance productivity and problem-solving capabilities among subcontractors.

Effective communication is also achieved through the use of modern tools and technology. Site managers leverage project management software, communication platforms, and mobile applications to streamline information flow. These tools enable real-time updates, document sharing, and efficient communication, leading to better coordination among subcontractors.

When challenges or conflicts arise, site managers take a proactive approach to address them promptly. They facilitate resolution discussions, mediating between different subcontractors if necessary. By resolving issues quickly and fairly, site managers maintain positive working relationships and prevent potential delays in the project.

In addition to formal communication channels, site managers maintain an approachable and accessible demeanor. They encourage subcontractors to approach them with questions or concerns at any time, fostering trust and openness. This accessibility helps in resolving issues swiftly and avoids miscommunication.

Moreover, site managers actively seek feedback from subcontractors about their experiences and suggestions for improvement. By valuing subcontractors' input, site managers create an atmosphere of mutual respect and demonstrate their commitment to continuous improvement.

It is essential for site managers to recognize and acknowledge the expertise and contributions of subcontractors. Showing appreciation for their efforts and celebrating their successes reinforces positive working relationships and motivates subcontractors to perform at their best.

Handling Payment Requests, Changes, and Disputes

In the realm of construction management, handling payment requests, changes, and disputes is a critical aspect of ensuring smooth project execution and maintaining positive relationships with subcontractors. Site managers play a pivotal role in managing these financial matters, as they directly impact the project's progress, budget, and overall success.

Payment Requests

Managing payment requests is a fundamental responsibility of site managers. When subcontractors complete their designated tasks, they submit payment requests for the work done. As site managers, it is essential to carefully review these requests to ensure that the work has been accurately and satisfactorily completed. Verification of the completion of milestones and adherence to quality standards are crucial steps in the evaluation process. Additionally, site managers must verify that all necessary documentation, such as invoices and receipts, accompanies the payment requests. Promptly processing valid payment requests is not only a contractual obligation but also helps maintain the subcontractors' confidence and motivation to deliver their best work.

Changes and Change Orders

Construction projects are inherently dynamic, and changes are not uncommon. Changes can arise due to unforeseen circumstances, adjustments in client requirements, or modifications to the project scope. Site managers must work closely with both the main contractor and subcontractors to assess the impact of these changes on the project's schedule, budget, and scope of work. Negotiating and implementing fair and reasonable change orders is vital to prevent misunderstandings and disputes. It is crucial to communicate changes effectively to subcontractors, ensuring that they have a clear understanding of the adjustments and the corresponding adjustments to their scope of work.

Dispute Resolution

Disagreements and disputes are sometimes unavoidable in construction projects. When disputes arise between the main contractor and subcontractors, site managers step in as mediators to facilitate constructive discussions and find amicable solutions. Effectively managing disputes requires careful review of the issue at hand, a thorough understanding of contractual agreements, and an unbiased approach to ensure fairness. Timely resolution of disputes helps maintain a harmonious project environment, fosters collaboration, and prevents delays in project completion.

Transparency and Documentation

To mitigate potential conflicts and misunderstandings, site managers must maintain thorough and accurate documentation of all financial transactions and agreements. This includes detailed records of payment requests, change orders, and any modifications to the original scope of work. Transparent processes and documentation enhance accountability, provide clarity to all stakeholders, and help build trust among team members.

Fair Payment Practices

Adhering to fair payment practices is not only a professional obligation but also a key element in sustaining positive relationships with subcontractors. Site managers must ensure that payment schedules and terms agreed upon in contracts are followed diligently. Timely payments to subcontractors are essential for their financial stability and continued dedication to the project. Consistency in payment practices reflects positively on the main contractor's reputation and attracts skilled subcontractors for future projects.

Continuous Communication

Effective communication is the foundation of successful financial management in construction projects. Site managers must establish and maintain regular communication channels with subcontractors to keep them informed about payment schedules, any changes in scope, and potential disputes. Open and transparent communication fosters trust, encourages collaboration, and enables swift resolution of issues. Regular updates and proactive discussions with subcontractors also help prevent misunderstandings and promote a shared sense of responsibility for the project's success.

Handling payment requests, changes, and disputes is a multifaceted responsibility for site managers. By demonstrating fairness, transparency, and effective communication, site managers can effectively manage financial matters and maintain harmonious relationships with subcontractors, ultimately contributing to the successful completion of construction projects.

Resolving Disputes and Maintaining Strong Relationships

In the dynamic and high-pressure environment of construction projects, disputes are inevitable. As a site manager, one of the most crucial skills is the ability to address conflicts promptly and effectively to ensure the smooth progression of the project and preserve strong relationships with all stakeholders involved. Resolving disputes in a fair and respectful manner not only minimizes disruptions but also fosters a collaborative atmosphere that can lead to improved project outcomes.

Active Listening and Understanding

When disputes arise, it is essential for site managers to actively listen to the concerns and perspectives of all parties involved. Taking the time to understand each party's viewpoint, interests, and underlying motivations lays the groundwork for finding common ground and reaching a satisfactory resolution. Demonstrating empathy and respect during discussions can help de-escalate tensions and encourage open dialogue.

Mediation and Facilitation

Site managers often act as mediators or facilitators in dispute resolution. Mediation involves bringing all parties together to work through the issues with the guidance of a neutral third party (the site manager). The site manager's role is to facilitate productive conversations, identify points of agreement, and explore potential solutions that address the interests of both sides. By creating a collaborative atmosphere, mediation can lead to win-win outcomes that preserve relationships and project progress.

Objective Assessment of Facts and Evidence

Resolving disputes requires a thorough and objective assessment of the facts and evidence at hand. Site managers must diligently review documentation, contractual agreements, and any other relevant information to gain a clear understanding of the situation. Objectively analyzing the information helps in making informed decisions that are fair and equitable.

Clear Communication and Timely Responses
Effective communication is vital in dispute resolution. Site managers must ensure that all parties are kept informed of the progress of the discussions, any decisions made, and the expected next steps. Timely responses to queries and concerns demonstrate professionalism and dedication to resolving the issue promptly. Clarity in communication also helps avoid misunderstandings and misconceptions that can exacerbate conflicts.

Seeking Win-Win Solutions
In resolving disputes, site managers should prioritize win-win solutions whenever possible. Win-win solutions are those that benefit all parties involved, leading to a sense of satisfaction and collaboration. It may involve finding creative alternatives or compromises that address the core interests of each party without compromising the project's overall objectives.

Preserving Professional Relationships
While disputes can be challenging, site managers must remain focused on preserving professional relationships. By handling disputes in a respectful and constructive manner, site managers can enhance trust and maintain the long-term cooperation of contractors, subcontractors, and other stakeholders. Healthy working relationships contribute to a positive work culture and increase the likelihood of successful future collaborations.

Implementing Lessons Learned
Dispute resolution also offers an opportunity for learning and

improvement. Site managers should take the time to reflect on the root causes of disputes and identify areas where processes can be strengthened to prevent similar issues in the future. Implementing lessons learned contributes to continuous improvement and enhances the overall efficiency of the project.

Resolving disputes and maintaining strong relationships is an essential aspect of effective site management. By employing active listening, mediation, and clear communication, site managers can navigate conflicts with professionalism and tact. Prioritizing win-win solutions and preserving professional relationships will contribute to the successful completion of construction projects and a positive reputation within the industry.

Handling Contractual Disputes and Claims

Contractual disputes and claims are inevitable in the construction industry and can arise due to a myriad of factors such as delays, changes in scope, payment issues, or disagreements over contract interpretation. As a site manager, it is essential to handle these disputes with utmost professionalism and efficiency to ensure smooth project execution and maintain positive relationships with stakeholders.

Thorough Contract Review: The first step in effectively addressing contractual disputes and claims is conducting a thorough review of the project contracts. Site managers should have a clear understanding of the terms and conditions, payment schedules, deliverables, and dispute resolution mechanisms outlined in the contracts. This knowledge will be crucial in presenting a well-informed and cohesive case during dispute resolution.

Open and Transparent Communication: Communication is a key element in managing disputes and claims. Site managers must foster open and transparent communication with all parties involved. Timely and clear communication can help prevent misunderstandings and enable parties to address issues proactively before they escalate into disputes.

Alternative Dispute Resolution (ADR): Resorting to litigation should be a last resort, as it can be time-consuming and costly. Site managers should explore alternative dispute resolution methods, such as mediation or arbitration, to facilitate a more amicable and efficient resolution. ADR can provide a neutral ground for parties to discuss and resolve their differences in a non-adversarial setting.

Documentation and Evidence Collection: Accurate and comprehensive documentation is essential in handling contractual disputes and claims. Site managers should maintain detailed records of all project-related activities, including contracts, change orders, progress reports, meeting minutes, and correspondence. These documents can serve as evidence to support the site manager's position during dispute resolution.

Professional Legal Assistance: In complex disputes with significant implications, seeking legal advice from construction law experts is advisable. Construction attorneys can offer valuable insights into contract interpretation, assess potential liabilities, and guide site managers in navigating the legal complexities of dispute resolution.

Cost-Benefit Analysis: Before deciding on a dispute resolution strategy, site managers should conduct a cost-benefit analysis. Evaluating the potential costs and risks associated with each approach can help site managers make informed decisions and select the most effective and efficient resolution method.

Focus on Solutions, Not Blame: When handling disputes, it is essential to focus on finding practical solutions rather than assigning blame. A collaborative and problem-solving approach can lead to quicker resolutions and help preserve relationships among project stakeholders.

Learn from Disputes and Claims: Each dispute or claim presents an opportunity for learning and improvement. Site managers should conduct post-mortems to identify the root causes and lessons learned from the experience. Implementing corrective actions based on these insights can help prevent similar issues in future projects.

Effective handling of contractual disputes and claims is crucial for site managers to ensure project success and maintain harmonious relationships with stakeholders. By conducting thorough contract reviews, fostering transparent communication, considering alternative dispute resolution methods, and seeking legal advice, when necessary, site managers can navigate these challenges with professionalism and expertise. Prioritizing problem-solving and continuous improvement will contribute to the efficient resolution of disputes, promoting successful project outcomes and enhancing the site manager's reputation in the construction industry.

Implementing Conflict Resolution Strategies

Conflict is a natural and unavoidable aspect of any construction project, given the complex interactions between various stakeholders and the high-pressure nature of the industry. As a site manager, it is crucial to proactively address conflicts and implement effective strategies to ensure a harmonious work environment and smooth project execution. Here are some key strategies for conflict resolution:

1. Early Identification and Intervention: Site managers should be vigilant in identifying signs of potential conflicts early on. This involves actively listening to team members, observing interactions, and addressing any emerging issues promptly. Early intervention can prevent conflicts from escalating and becoming more challenging to resolve.

2. Active Listening and Empathy: Effective conflict resolution starts with active listening and demonstrating empathy towards all parties involved. By understanding the concerns and perspectives of each individual, site managers can build trust and create a safe space for open dialogue.

3. Open Communication Channels: Establishing clear and open communication channels is vital for addressing conflicts. Site managers should encourage regular team meetings, one-on-one discussions, and anonymous feedback mechanisms to enable team members to express their thoughts and concerns freely.

4. Mediation and Facilitation: When conflicts arise, site managers can act as mediators or facilitators to help parties involved find common ground and reach a resolution. Mediation allows for a neutral third-party to guide the discussions, helping parties focus on problem-solving rather than escalating emotions.

5. Conflict Resolution Training: Providing conflict resolution training to team members can equip them with the necessary skills to manage conflicts constructively. Training sessions can cover active listening, effective communication, negotiation techniques, and de-escalation strategies.

6. Win-Win Solutions: Encouraging a collaborative approach to conflict resolution, site managers should strive for win-win solutions. These solutions aim to address the needs and interests of all parties, fostering a sense of cooperation and shared success.

7. Clear Policies and Procedures: Having clear and well-defined policies and procedures for conflict resolution is essential. This can include formal processes for escalating issues, handling grievances, and involving higher management when necessary.

8. Focus on the Issue, Not Personalities: Conflict resolution should focus on addressing the specific issue at hand rather than personal attacks. Site managers should steer discussions towards the root cause of the conflict and avoid getting entangled in personal disagreements.

9. Follow-Up and Monitoring: After implementing conflict resolution strategies, site managers should conduct follow-ups to ensure that the resolution remains effective. Regular monitoring can identify any recurring conflicts or issues and allow for further improvements in conflict management processes.

10. Learn from Conflicts: Each conflict presents an opportunity for learning and growth. Site managers should conduct post-conflict reviews to identify the underlying causes and lessons learned. These insights can help refine conflict resolution strategies for future projects.

Conflict resolution is an essential skill for site managers to maintain a positive and productive work environment in the construction industry. By employing proactive measures, active listening, and empathy, implementing mediation techniques, and promoting win-win solutions, site managers can effectively address conflicts and build strong, collaborative teams that contribute to successful project outcomes.

Nurturing Long-Term Collaboration with Subcontractors

Building and maintaining strong relationships with subcontractors is integral to the success of construction projects. As a site manager, fostering long-term collaboration with subcontractors can lead to improved project outcomes, increased efficiency, and enhanced trust among all parties involved. Here are some key strategies to nurture and strengthen these essential partnerships:

1. Selecting the Right Subcontractors: It all begins with the careful selection of subcontractors who align with the project's requirements and values. Site managers should conduct thorough evaluations of subcontractor qualifications, past performance, and reputation in the industry. Choosing reliable and experienced subcontractors sets the foundation for a successful collaboration.

2. Transparent Communication: Effective communication forms the backbone of any successful collaboration. Site managers should establish clear and transparent lines of communication with subcontractors from the outset. Regular meetings, progress updates, and open discussions about project expectations, timelines, and any potential challenges can foster trust and alignment.

3. Fair and Timely Payment: Ensuring fair and timely payment to subcontractors is crucial for maintaining a positive relationship. Site managers should adhere to agreed-upon payment terms and promptly address any payment-related concerns or delays.

4. Collaborative Decision-Making: Involving subcontractors in decision-making processes can create a sense of ownership and investment in the project. Soliciting their input and expertise can lead to innovative solutions and promote a collaborative working environment.

5. Acknowledging and Appreciating Contributions: Recognizing the efforts and contributions of subcontractors is essential for building a strong partnership. A simple gesture of appreciation can go a long way in motivating subcontractors to perform at their best.

6. Resolving Conflicts Amicably: Conflicts can arise in any project, but how they are handled defines the strength of the collaboration. Site managers should approach conflicts with a problem-solving mindset, seeking win-win solutions that satisfy the interests of both parties.

7. Providing Opportunities for Growth: Offering repeat business and future project opportunities to subcontractors who consistently deliver high-quality work can incentivize them to maintain a long-term commitment to the collaboration.

8. Regular Performance Reviews: Conducting regular performance reviews with subcontractors can help assess their strengths and identify areas for improvement. Providing constructive feedback can lead to continuous improvement and foster a sense of partnership in achieving project goals.

9. Networking and Industry Involvement: Encouraging subcontractors to participate in industry events and professional associations can expand their network and provide access to valuable resources. Site managers can also facilitate networking opportunities to strengthen their position in the industry.

10. Celebrating Success Together: Celebrating project milestones and successes as a team reinforces a sense of camaraderie and achievement. Site managers can organize events or ceremonies to recognize the collective effort of all stakeholders, including subcontractors.

CHAPTER 7:
RESOURCE AND EQUIPMENT MANAGEMENT

E ffective resource and equipment management is vital for the smooth execution of construction projects. As a site manager, it is your responsibility to optimize the utilization of labor, materials, and machinery to ensure efficient operations and timely project delivery. This chapter delves into various strategies and techniques to manage resources and equipment effectively, enhancing productivity and minimizing delays.

Efficient Utilization of Labor and Equipment Resources

Efficient utilization of labor and equipment resources is a cornerstone of successful construction site management. As a site manager, it is crucial to carefully plan and allocate human resources and machinery to maximize productivity and meet project timelines. By understanding the specific skill sets of your workforce and the capabilities of your equipment, you can create a well-coordinated and productive work environment.

To optimize labor resources, start by conducting a thorough assessment of the required skills and expertise for each task within the project. By matching the right individuals to the right tasks, you can ensure that work is carried out efficiently and to the highest standard. Regularly monitor the performance of your workforce and provide them with the necessary training and support to enhance their skills and productivity.

Similarly, efficient equipment management involves meticulous planning and scheduling. Identify the types of machinery needed for various project phases and acquire reliable and well-maintained equipment from trusted suppliers. Regular maintenance and inspections are essential to prevent breakdowns and downtime, which can significantly impact project progress.

Implementing a well-structured labor and equipment scheduling system is vital. This system should consider factors such as task priorities, equipment availability, and resource allocation based on project timelines. By maintaining a balanced workload and avoiding resource bottlenecks, you can minimize delays and optimize project efficiency.

Embracing modern technologies and construction management software can aid in resource optimization. Utilize project management tools to track labor hours, monitor equipment usage, and identify potential areas of improvement. Real-time data and analytics can offer valuable insights for informed decision-making and resource allocation adjustments.

By prioritizing efficient resource utilization, site managers can significantly contribute to the overall success of construction projects, meeting deadlines, and delivering projects within budget while maintaining a high standard of quality.

Planning Labor Requirements and Workforce Allocation

Planning labor requirements and workforce allocation is a critical aspect of resource and equipment management for site managers. A well-thought-out labor plan ensures that the right number of skilled workers is available at each stage of the construction project, preventing delays and ensuring optimal productivity.

Start by conducting a detailed assessment of the project's scope and timeline. Break down the project into specific tasks and identify the skill sets required for each activity. Consider the complexity of the tasks, the estimated time for completion, and any potential overlaps or dependencies between different activities.

Once you have a clear understanding of the labor requirements, assess the available workforce. Evaluate the skills, experience, and qualifications of your existing team members to determine if they align with the project's demands. Identify any gaps in expertise and plan for recruitment or training to fill those gaps.

Efficient workforce allocation involves balancing the workload and ensuring that each worker is assigned tasks that align with their strengths. Avoid overburdening certain team members while underutilizing others. Utilize a collaborative approach to involve workers in the allocation process, taking their preferences and expertise into account.

Maintain flexibility in your labor plan to accommodate unforeseen circumstances or changes in project requirements. Develop contingency plans for possible labor shortages or surpluses and be prepared to adjust the workforce allocation as needed.

Consider factors such as shift schedules and overtime requirements to optimize labor productivity while adhering to labor regulations and safety standards. A well-planned labor schedule ensures that the construction site remains operational without compromising the well-being of the workers.

Additionally, effective communication is key to successful labor planning and workforce allocation. Keep all team members informed about the project's progress, any changes in schedules, and their individual roles and responsibilities. Regularly update workers on project milestones and goals to maintain motivation and alignment with project objectives.

By carefully planning labor requirements and workforce allocation, site managers can enhance efficiency, minimize downtime, and ensure the timely completion of construction projects while fostering a positive work environment for the team.

Optimizing Equipment Usage and Scheduling

Optimizing equipment usage and scheduling is crucial for effective resource and equipment management on construction sites. Construction equipment represents a significant investment, and its efficient use can lead to cost savings, increased productivity, and reduced project timelines.

Start by conducting a comprehensive inventory of all construction equipment available for the project. Assess the capabilities and capacities of each machine, considering factors such as load-bearing capacity, power output, and fuel efficiency. Identify any redundancies or gaps in the equipment fleet and make necessary adjustments to ensure that the right equipment is available for each task.

Implement a centralized system for equipment scheduling and tracking. Utilize technology, such as construction management software or mobile applications, to manage equipment reservations, usage logs, and maintenance schedules. A well-organized system enables site managers to allocate equipment to specific tasks and avoid conflicts in scheduling.

Adopt a preventive maintenance program to ensure that all equipment remains in optimal working condition throughout the project. Regular inspections, servicing, and repairs help prevent unexpected breakdowns and minimize downtime. Maintain detailed maintenance records and address any issues promptly to extend the lifespan of the equipment and avoid costly repairs.

Consider the sequencing of tasks and plan equipment usage accordingly. Optimize the equipment schedule to minimize idle time and ensure that each machine is utilized to its full potential. Identify opportunities for equipment sharing between tasks or work areas to further enhance efficiency.

Promote effective communication and coordination between equipment operators and other team members. Clear communication channels help avoid delays caused by miscommunication or misunderstandings. Ensure that operators are aware of their responsibilities and are trained to handle the equipment safely and efficiently.

Monitor equipment usage and performance regularly to identify areas for improvement. Analyze data on fuel consumption, operating hours, and productivity to make informed decisions about equipment usage and allocation. Use performance metrics to evaluate the effectiveness of the equipment management strategy and make adjustments as needed.

Lastly, prioritize safety in equipment operations. Provide comprehensive training to equipment operators on safety protocols and best practices. Regularly inspect the equipment for safety compliance and address any safety concerns immediately. A safe work environment not only protects workers but also prevents accidents that can lead to costly delays and damage to equipment.

By optimizing equipment usage and scheduling, site managers

can streamline operations, increase productivity, and ensure that construction projects are completed efficiently and within budget. Effective equipment management is a key component of successful construction site operations.

Balancing Resource Allocation with Project Demands

Balancing resource allocation with project demands is a critical aspect of effective resource and equipment management on construction sites. It involves ensuring that the right resources, including labor, materials, and equipment, are available in the right quantities and at the right times to meet project requirements and deadlines.

Start by conducting a thorough analysis of the project scope, timeline, and specific resource needs. Understand the various tasks and activities that need to be completed and the corresponding resource requirements for each phase of the project. Consider factors such as the complexity of tasks, the skill level of the workforce, and the availability of materials and equipment.

Develop a detailed resource plan that outlines the allocation of resources throughout the project's lifecycle. This plan should include a schedule for when and where resources will be needed, as well as contingency plans for unexpected changes or challenges.

Communicate with project stakeholders, including clients, subcontractors, and suppliers, to align resource allocation with project milestones and deliverables. Collaborative communication ensures that everyone involved is aware of resource needs and can plan accordingly.

Utilize construction management software or other technology tools to monitor resource allocation and utilization in real-time. These tools can help track labor hours, material inventory, and equipment usage, allowing site managers to identify potential bottlenecks or areas where resources may be underutilized.

Implement a flexible approach to resource allocation, allowing for adjustments as the project progresses. Changes in project scope, unexpected delays, or unforeseen challenges may require reallocating resources to ensure project efficiency and success.

Consider the skill set and experience of the workforce when allocating labor resources. Assign tasks to individuals with the right expertise and training to optimize productivity and minimize errors.

Regularly review and analyze resource allocation data to identify opportunities for improvement. Look for trends or patterns that may indicate areas of resource inefficiency and implement strategies to address them.

Strive for a balanced allocation of resources to avoid overloading certain areas while neglecting others. Keeping a balance ensures that all project aspects receive the attention and resources they need for successful completion.

Lastly, maintain open communication with the project team and stakeholders throughout the project. Regularly update them on resource allocation decisions and involve them in resource planning discussions when appropriate.

By effectively balancing resource allocation with project demands, site managers can optimize project performance, enhance productivity, and achieve successful project outcomes. Adapting resource allocation as needed and staying proactive in resource management are key to overcoming challenges and maximizing project success.

Preventive Maintenance for Construction Machinery
Preventive maintenance for construction machinery is a proactive approach to ensure that equipment operates at its optimal performance levels and remains in good working condition throughout the project's duration. It involves regularly scheduled inspections, servicing, and repairs to

prevent breakdowns, minimize downtime, and extend the lifespan of the machinery.

To implement an effective preventive maintenance program, site managers should follow these key steps:

Equipment Inventory and Documentation: Create a comprehensive inventory of all construction machinery used on-site. Record essential information for each piece of equipment, including make, model, serial number, and purchase date. This documentation serves as a reference for maintenance schedules and helps track equipment history.

Manufacturer Guidelines: Refer to the manufacturer's guidelines and maintenance manuals for each piece of machinery. Manufacturers typically provide specific maintenance recommendations, including inspection intervals, lubrication requirements, and suggested replacement schedules.

Maintenance Schedule Development: Based on the manufacturer's guidelines and equipment usage, develop a maintenance schedule that includes regular inspections, lubrication, cleaning, and other preventive measures. Assign responsibilities for each task to qualified personnel.

Regular Inspections: Conduct routine inspections of construction machinery before and after each use. Inspect for signs of wear, damage, leaks, and loose components. Address any issues promptly to prevent further damage or safety hazards.

Lubrication and Fluid Checks: Ensure that all moving parts are adequately lubricated according to the manufacturer's recommendations. Regularly check fluid levels, such as engine oil, hydraulic fluid, coolant, and fuel, and top them up as needed.

Filter Replacement: Replace air, fuel, and oil filters at recommended intervals to maintain optimal engine performance and reduce the risk of contaminants damaging the machinery.

Cleaning and Maintenance of Attachments: Keep attachments and accessories clean and well-maintained to ensure proper functionality. Regularly inspect and service attachments such as buckets, blades, and forks.

Battery Maintenance: Check and maintain batteries regularly, including cleaning terminals and checking electrolyte levels. Replace batteries as needed to avoid unexpected breakdowns.

Tire Maintenance: Inspect tires for wear and damage regularly. Ensure proper inflation and alignment to optimize equipment performance and extend tire life.

Keeping Records: Maintain detailed records of all preventive maintenance activities, including inspection results, repairs, and parts replacements. These records help track equipment history and inform future maintenance decisions.

Implementing a preventive maintenance program helps site managers avoid costly breakdowns and delays, reduce repair expenses, and enhance overall construction site safety. By prioritizing preventive maintenance, site managers can ensure that their construction machinery remains reliable and efficient, contributing to the successful completion of projects on time and within budget.

Developing a Preventive Maintenance Program

Developing a comprehensive preventive maintenance program is crucial for effective site management. A well-designed program ensures that construction machinery and equipment operate at their peak performance, reduces downtime, prevents costly breakdowns, and enhances safety on the construction site. Here are the essential steps to develop an effective preventive maintenance program:

Equipment Inventory and Assessment: Begin by creating a detailed inventory of all construction machinery and equipment used on-site. Assess each item's criticality and importance to the project to prioritize maintenance efforts.

Manufacturer's Guidelines: Consult the manufacturer's manuals and guidelines for each piece of equipment. These documents provide valuable information on maintenance schedules, recommended parts, lubrication requirements, and inspection procedures.

Maintenance Schedule and Tasks: Develop a comprehensive maintenance schedule that outlines regular maintenance tasks, such as inspections, lubrication, cleaning, and calibration. Assign specific responsibilities to qualified personnel for each task.

Inspection Checklists: Create standardized inspection checklists for different types of equipment. These checklists should cover all critical components and potential failure points to ensure a thorough inspection.

Lubrication and Fluid Management: Implement a systematic approach to lubricate equipment according to the manufacturer's recommendations. Monitor fluid levels regularly and adhere to proper disposal procedures for used fluids.

Calibration and Testing: Periodically calibrate equipment that requires precise measurements, such as surveying instruments or electronic tools. Testing ensures accuracy and reliability during use.

Replacement Parts Inventory: Establish an inventory of frequently used replacement parts, such as filters, belts, hoses, and batteries. Ensure that spare parts are readily available for prompt repairs.

Training and Certification: Provide training and certification programs for maintenance personnel to ensure they have the necessary skills and knowledge to carry out preventive maintenance tasks effectively.

Record Keeping: Maintain detailed records of all preventive maintenance activities, including inspection results, repairs, parts replacements, and calibration records. These records aid in tracking equipment performance and identifying trends.

Continuous Improvement: Regularly review the preventive maintenance program's effectiveness and identify areas for improvement. Seek feedback from maintenance personnel and equipment operators to refine the program as needed.

Compliance with Regulations: Ensure that the preventive maintenance program complies with all relevant industry regulations and safety standards.

Emergency Response Plan: Develop an emergency response plan to address unforeseen breakdowns or critical equipment failures promptly. Define procedures for rapid response, troubleshooting, and repair.

Through a well-structured preventive maintenance program, site managers can minimize equipment downtime, reduce maintenance costs, enhance equipment reliability, and ultimately improve the overall productivity and efficiency of the construction site. A proactive approach to equipment maintenance not only saves time and resources but also fosters a safe working environment, which is vital for the success of any construction project.

Scheduling Regular Equipment Inspections and Servicing

Scheduling regular equipment inspections and servicing is a fundamental aspect of effective resource and equipment management on construction sites. By adhering to a well-planned inspection and servicing schedule, site managers can ensure that construction machinery operates optimally, identify potential issues before they escalate, and extend the lifespan of the equipment. Here are the key steps in scheduling regular equipment inspections and servicing:

Establish Inspection Intervals: Determine the appropriate inspection intervals for each type of equipment based on manufacturer recommendations, usage frequency, and the severity of operational conditions. Some equipment may require daily inspections, while others may be inspected weekly, monthly, or after a certain number of operating hours.

Create an Inspection Calendar: Develop a comprehensive inspection calendar that outlines the inspection dates and the specific equipment to be inspected on each occasion. Use digital tools or software to schedule reminders and notifications for upcoming inspections.

Pre-Operation Checks: Prioritize pre-operation checks before using equipment each day. Operators should perform routine checks to verify fluid levels, tire pressure, brakes, lights, and other critical components to ensure safe and reliable operation.

Standardized Inspection Procedures: Standardize the inspection procedures for each type of equipment. Create detailed checklists that cover all essential components, including engine, hydraulics, electrical systems, safety features, and any attachments.

Qualified Inspection Personnel: Assign qualified and trained personnel to conduct inspections. These individuals should have a solid understanding of equipment operation, maintenance requirements, and the ability to identify potential issues.

Servicing and Maintenance: Schedule regular servicing and maintenance based on the equipment's operating hours or the manufacturer's recommendations. This includes oil changes, filter replacements, belt adjustments, and other routine maintenance tasks.

Records and Documentation: Maintain detailed records of all inspections, servicing, and maintenance activities. Record any identified issues, the actions taken for rectification, and the date of completion. Proper documentation aids in tracking equipment history and compliance with warranties.

Equipment Tracking System: Utilize an equipment tracking system to monitor the inspection and servicing history of each machine. This system facilitates efficient equipment management and helps to avoid missed inspections.

Training and Awareness: Conduct training sessions to educate equipment operators on the importance of regular inspections and servicing. Encourage a culture of proactive equipment care and safety awareness among all site personnel.

Responsive Maintenance: Promptly address any issues or abnormalities discovered during inspections. Responding quickly to equipment problems can prevent costly breakdowns and ensure smooth project progress.

Through a well-organized schedule of regular equipment inspections and servicing, site managers can enhance equipment reliability, promote safety, and streamline construction operations. This proactive approach to equipment management contributes significantly to the overall success of construction projects and supports the efficient use of resources on-site.

Extending Equipment Lifespan and Reducing Downtime

Efficiently managing construction equipment is a critical responsibility for site managers, as it directly impacts project timelines, costs, and overall productivity. Extending the lifespan of equipment and minimizing downtime are essential goals to achieve successful project outcomes.

Regular Maintenance and Inspections: Regular maintenance and inspections are fundamental in extending equipment lifespan. Implement a preventive maintenance program that includes routine checks, servicing, and component replacements based on manufacturer recommendations or operating hours. Timely identification and resolution of issues can prevent minor problems from developing into major breakdowns, reducing costly downtime.

Operator Training and Skill Development: Invest in comprehensive training for equipment operators to ensure they understand proper usage, care, and safety measures. Skilled operators can prevent unnecessary wear and tear on machinery, leading to extended equipment life.

Utilizing Equipment Tracking Systems: Employing advanced equipment tracking systems allows site managers to monitor equipment usage, operating hours, maintenance history, and upcoming servicing needs. This data-driven approach enables proactive decision-making and reduces the risk of unexpected breakdowns.

Optimizing Equipment Utilization: Assess equipment requirements and optimize their allocation based on project demands. Overloading or underutilizing machinery can lead to premature wear and inefficiencies. Right-sizing equipment usage can contribute to longer equipment lifespan.

Environmental Considerations: Take environmental factors into account when operating equipment. Properly clean and store equipment after use, especially in harsh weather conditions, to minimize rust and corrosion. Implementing protective measures can significantly extend the equipment's lifespan.

Regular Equipment Training for Staff: Conduct regular

equipment training for all site staff, not just operators. Ensuring that everyone handling equipment is familiar with its proper usage and maintenance practices can prevent avoidable damage and prolong equipment life.

Prompt Repairs and Upgrades: Address equipment issues promptly and invest in necessary repairs or upgrades. Delaying repairs can exacerbate problems and lead to more extensive damage, while timely fixes can keep equipment functioning optimally.

Equipment Storage and Shelter: Store equipment in a secure and covered area when not in use. Sheltering equipment from the elements can prevent weather-related damage and protect sensitive components, promoting longer lifespan.

Monitoring Equipment Performance: Monitor equipment performance regularly to identify patterns of wear or inefficiency. Address recurring problems proactively, and consider upgrading to more durable models or incorporating modern technologies where applicable.

Collaboration with Equipment Suppliers: Develop strong relationships with equipment suppliers and manufacturers. Engaging with them can provide valuable insights on maintenance best practices and access to genuine spare parts, ensuring the equipment operates at its best for longer periods.

Through these practices, site managers can effectively extend the lifespan of construction equipment and minimize downtime. These efforts not only contribute to cost savings but also enhance overall project efficiency, safety, and success. An investment in equipment management ultimately results in a higher return on investment and a more sustainable construction operation.

Controlling Equipment Costs and Downtime

Efficient management of equipment costs and downtime is a

crucial aspect of successful construction project management. Aspiring and current site managers must implement effective strategies to optimize equipment utilization, minimize expenses, and reduce unproductive hours. Through proactive measures, site managers can ensure that the project stays on budget, meets deadlines, and achieves its objectives.

One of the key steps in controlling equipment costs is maintaining a comprehensive inventory. Keeping track of all equipment, including purchase dates, maintenance records, and depreciation values, enables site managers to make informed decisions about equipment replacements, upgrades, or rentals. A well-defined equipment procurement strategy is also essential. By carefully considering project-specific equipment needs and exploring rental or leasing options, site managers can make cost-effective decisions and minimize unnecessary expenses.

To reduce downtime and enhance equipment performance, site managers should prioritize preventive maintenance. This shift from reactive to predictive maintenance involves using technology and data analytics to identify potential equipment failures before they occur. This not only minimizes downtime but also saves on costly unplanned repairs. Regularly monitoring equipment utilization helps identify underutilized machinery, allowing site managers to optimize equipment allocation and avoid unnecessary costs.

When leasing equipment, site managers must manage lease agreements effectively to prevent penalties for extended use or damage. Strategically utilizing equipment rentals for short-term or specialized needs can be more cost-efficient than purchasing expensive machinery with limited usage. Additionally, maintaining a comprehensive inventory of spare parts and consumables ensures that critical components are readily available, reducing repair time and associated costs.

Investing in training and skill development for equipment

operators and maintenance personnel is essential for improving efficiency and minimizing equipment-related incidents. By analyzing equipment performance and contractor performance against established benchmarks, site managers can identify inefficiencies and address them promptly. Proper budget allocation is also crucial in covering equipment costs and planning for maintenance and repair expenses.

Monitoring Equipment Expenses and Budgets

Monitoring equipment expenses and budgets is an essential aspect of effective resource management for site managers. Aspiring and current site managers must have a clear understanding of the financial aspects related to equipment procurement, maintenance, and operation. Regularly tracking and analyzing equipment expenses allows site managers to make informed decisions, optimize resource allocation, and ensure that projects remain within budgetary constraints.

To monitor equipment expenses, site managers should maintain accurate and up-to-date records of all costs associated with equipment. This includes initial purchase or lease costs, transportation expenses, installation fees, and ongoing maintenance expenditures. By organizing this data systematically, site managers can identify patterns and trends in equipment-related costs, enabling them to take proactive measures to control expenses.

Creating a detailed equipment budget at the outset of a project is crucial. The budget should account for all anticipated equipment-related costs throughout the project's duration. This includes not only direct costs but also indirect expenses such as insurance, licensing fees, and training for equipment operators. Site managers should regularly review the budget and make adjustments as necessary to accommodate any unforeseen expenses or changes in project scope.

By comparing actual expenses against the budget, site managers can identify any discrepancies and take corrective actions to avoid cost overruns. Regular financial reporting and analysis provide valuable insights into the financial health of the project and help site managers identify areas where cost-saving measures can be implemented.

In addition to monitoring expenses, site managers must also track equipment utilization and downtime. By assessing how frequently equipment is being used and how much downtime is occurring, site managers can identify opportunities for improving equipment efficiency and reducing unnecessary costs.

Implementing technology solutions, such as equipment tracking software and telemetry systems, can streamline the monitoring process and provide real-time data on equipment performance. This data-driven approach enables site managers to make data-based decisions, optimize equipment usage, and address maintenance issues promptly to minimize downtime.

Establishing open lines of communication with the equipment operators and maintenance teams is essential. They can provide valuable insights into equipment performance and identify any potential issues that may affect costs. Regular meetings and feedback sessions allow site managers to address concerns promptly and collaboratively find solutions.

Identifying Cost-Saving Opportunities in Equipment Management
Identifying cost-saving opportunities in equipment management is a critical task for site managers. Aspiring and current site managers need to be vigilant in seeking ways to optimize equipment usage, reduce wasteful expenditures, and enhance overall operational efficiency. By identifying and implementing cost-saving measures, site managers can contribute significantly to the financial success of construction projects.

One effective approach to identifying cost-saving opportunities is conducting a thorough equipment assessment. Site managers should review the current equipment inventory, its age, condition, and performance history. By analyzing this data, they can determine whether certain equipment is becoming more costly to maintain and operate due to age or wear and tear. This assessment helps in making informed decisions about whether to repair, replace, or lease new equipment, ensuring the most cost-effective choice is made.

Implementing preventive maintenance programs can also lead to significant cost savings. Regularly scheduled maintenance and inspections can prevent costly breakdowns and reduce the need for major repairs. Site managers can work closely with maintenance teams to establish maintenance schedules and ensure compliance with manufacturer recommendations, thereby prolonging equipment lifespan and optimizing its performance.

Another cost-saving opportunity lies in exploring equipment rental options. Instead of purchasing expensive equipment that may only be needed for a specific project phase, site managers can opt to rent equipment as required. Renting equipment allows for more flexibility in managing resources and avoids long-term maintenance costs and storage expenses.

Furthermore, leveraging technology and telematics can offer cost-saving advantages. Monitoring equipment usage and performance through telematics systems enables site managers to detect inefficiencies and address operational issues promptly. These technologies provide valuable data on fuel consumption, idle time, and equipment productivity, allowing for data-driven decisions and potential operational improvements.

Site managers should also consider adopting eco-friendly and energy-efficient equipment. Modern equipment designed with energy-saving features can result in reduced fuel consumption and lower maintenance expenses. Not only do these initiatives contribute to cost savings, but they also demonstrate a commitment to sustainability, which is increasingly important in the construction industry.

Additionally, fostering a culture of equipment accountability among the workforce can minimize misuse and abuse of equipment, leading to extended equipment life and reduced replacement costs. Proper training for equipment operators can enhance their efficiency and prevent unnecessary wear and tear.

Regularly benchmarking equipment-related costs against industry standards and best practices is another effective method of identifying potential cost-saving opportunities. This practice allows site managers to compare their equipment expenses with industry peers, ensuring they stay competitive and strive for continuous improvement.

Mitigating Equipment-Related Delays and Downtime

Mitigating equipment-related delays and downtime is of utmost importance for aspiring and current site managers. Unplanned equipment breakdowns and inefficiencies can lead to significant project delays, increased costs, and reduced productivity. Therefore, site managers must implement proactive strategies to minimize such disruptions and ensure smooth construction operations.

One crucial step in mitigating equipment-related delays is adhering to a strict preventive maintenance schedule. Regular and comprehensive maintenance checks can identify potential issues before they escalate into costly breakdowns. Site managers should work closely with maintenance teams to develop detailed maintenance plans that encompass all equipment on the construction site. This includes routine inspections, lubrication, parts replacement, and adherence to manufacturer recommendations. By keeping equipment in optimal condition, the risk of unexpected failures can be significantly reduced.

Furthermore, site managers must establish a robust inventory management system to track equipment usage and availability. Implementing a real-time tracking system can help monitor equipment location, usage hours, and upcoming maintenance needs. This ensures that equipment is utilized efficiently and that potential issues can be addressed promptly. Additionally, having a clear understanding of equipment availability allows site managers to plan construction activities better and allocate resources effectively.

Investing in high-quality and reliable equipment also plays a significant role in mitigating delays. While it may be tempting to opt for cheaper options, subpar equipment may be more prone to breakdowns and inefficiencies. By selecting reputable equipment suppliers and investing in well-maintained machinery, site managers can reduce the risk of downtime and unexpected delays.

Cross-training equipment operators can enhance workforce flexibility and reduce reliance on a single skilled operator. In cases where a specific equipment operator is unavailable due to unforeseen circumstances, having multiple operators who can handle various equipment ensures that operations can continue seamlessly.

Incorporating contingency plans and backup equipment options is another effective strategy to mitigate delays. By having backup equipment readily available, site managers can quickly address unexpected breakdowns and ensure that critical construction tasks continue without interruption.

Additionally, fostering open communication and collaboration among the project team is essential. Encouraging equipment operators and maintenance staff to report any issues or potential concerns promptly allows for quick resolution and minimizes the impact on project timelines.

Lastly, conducting regular performance evaluations of

both equipment and operators can help identify areas for improvement and optimize productivity. By reviewing equipment performance data and operator efficiency, site managers can identify trends and implement targeted training or adjustments to maximize equipment utilization.

CHAPTER 8: PROGRESS MONITORING AND REPORTING

This chapter focuses on the critical aspect of progress monitoring and reporting for aspiring and current site managers. It provides valuable insights into methodologies and tools that site managers can utilize to effectively track project advancement, identify potential issues, and communicate progress to stakeholders in a clear and concise manner. From setting clear performance indicators to conducting regular site inspections and utilizing project management software, this chapter equips site managers with essential techniques to ensure the success of construction projects. Additionally, it emphasizes the importance of risk assessment, client and stakeholder reporting, and the implementation of earned value management techniques to forecast project outcomes and make necessary adjustments. Overall, this chapter serves as a comprehensive guide for site managers to master the art of progress monitoring and reporting, contributing to the smooth execution of construction projects.

Tracking Construction Progress and Milestones

Tracking construction progress and milestones is a crucial responsibility for site managers to ensure the successful execution of construction projects. This process involves continuously monitoring and assessing the status of various activities to ensure they are progressing as planned and are in line with the project schedule and objectives.

One of the primary methods used for tracking construction progress is the development of a detailed work breakdown structure (WBS). The WBS breaks down the project into smaller, manageable tasks, each with its own set of deliverables and deadlines. By organizing the project into these smaller components, site managers can easily track the completion of each task and identify potential bottlenecks or delays.

Regular site inspections and daily log reports are essential tools for tracking progress on the construction site. Site managers should conduct routine inspections to visually assess the work being done, verify the quality of construction, and ensure that safety protocols are being followed. Daily log reports provide a written record of daily activities, workforce allocation, and any challenges faced, allowing site managers to identify trends and patterns in the project's progress.

Project management software also plays a significant role in tracking construction progress. These software solutions offer real-time data collection and analysis, enabling site managers to monitor task completion, resource allocation, and overall project performance. The visual representation of data through charts and graphs provides a clear and comprehensive overview of the project's status.

In addition to monitoring progress, site managers must also track project milestones. Milestones are significant events or achievements within the project timeline that serve as critical points of reference for progress evaluation. Examples of milestones include completing the foundation, reaching the halfway point of construction, or obtaining necessary permits.

Tracking construction progress and milestones allows site managers to identify potential issues and risks early on in the project lifecycle. By doing so, they can take proactive measures to address these challenges and prevent them from escalating into more significant problems that could lead to delays or cost overruns.

Regular progress meetings with the project team and stakeholders are another effective way to track construction progress. These meetings provide an opportunity to discuss completed tasks, upcoming milestones, and any roadblocks that may be hindering progress. It also fosters open communication and collaboration among all parties involved.

Additionally, site managers can use analytical techniques such as the critical path method (CPM) and earned value analysis (EVA) to gain deeper insights into the project's performance. CPM helps identify the longest sequence of tasks and highlights potential areas of delay, while EVA assesses project progress in terms of cost and schedule performance.

Tracking construction progress and milestones is an integral part of effective project management for site managers. Through a combination of diligent monitoring, regular reporting, and the use of modern tools and techniques, site managers can ensure that construction projects stay on track, meet established deadlines, and achieve the desired quality and safety standards.

Implementing Progress Tracking Systems and Tools
Implementing progress tracking systems and tools is a critical aspect of effective construction project management. In today's fast-paced and dynamic construction environment,

site managers must have access to real-time data and comprehensive insights to make informed decisions and keep projects on track. These tracking systems and tools play a pivotal role in streamlining project monitoring, facilitating communication, and ensuring successful project delivery.

One of the key benefits of implementing progress tracking systems is the ability to centralize project information. These systems provide a centralized platform where project data, updates, and documentation can be stored, organized, and accessed by authorized team members. This eliminates the need for scattered documents and spreadsheets, reducing the risk of miscommunication and data inconsistencies.

Progress tracking systems offer various features that enhance project oversight and control. These features include task scheduling, resource allocation, document management, cost tracking, and reporting capabilities. With these tools at their disposal, site managers can effectively plan, assign tasks, and monitor the progress of each activity in the construction project.

Real-time data is a significant advantage of modern progress tracking systems. Site managers can access up-to-date information on project status, task completion, and resource utilization from any location with internet access. This accessibility fosters efficient decision-making and allows site managers to address potential issues promptly before they escalate.

Many progress tracking tools also integrate with BIM and Internet of Things (IoT) technologies. The integration of BIM models allows site managers to visualize the project in a 3D representation, aiding in project planning and identification of potential clashes or conflicts. IoT devices can provide real-time data on equipment performance and resource usage, facilitating better resource management and scheduling.

Training and orientation are crucial components of successful progress tracking implementation. Site managers and team members must receive adequate training to ensure they can use the system efficiently and effectively. This training empowers the team to embrace the new technology and maximize its potential for project success.

Regular communication and collaboration are fostered by progress tracking systems. Team members can communicate updates, share documents, and raise concerns within the platform, promoting transparency and facilitating efficient collaboration among all stakeholders.

Continuous improvement is vital when implementing progress tracking systems and tools. Feedback from the project team is valuable in identifying areas for improvement and refining the system to better suit the project's specific needs.

Reporting Milestones and Achievements to Stakeholders

Reporting milestones and achievements to stakeholders is a critical aspect of effective project management. Stakeholders, including clients, investors, senior management, and other key decision-makers, rely on timely and accurate updates to assess the project's progress, identify potential risks, and make informed decisions. Site managers play a pivotal role in communicating project status and ensuring stakeholders are kept informed throughout the construction process.

The reporting process begins with setting clear milestones and objectives at the outset of the project. These milestones serve as benchmarks to measure progress and achievements at various stages of the project. Site managers must work closely with project teams to establish achievable milestones and define the criteria for success.

Regular reporting is essential to keep stakeholders informed. Depending on the project's scale and complexity, reporting frequency may vary, but it is typically done on a weekly, bi-weekly, or monthly basis. The reports should be concise, yet comprehensive, providing key performance indicators (KPIs), project status updates, and progress against established milestones.

To ensure effective reporting, site managers must use a consistent and standardized format that aligns with stakeholders' expectations. The reports should be well-structured, visually appealing, and easy to understand. Including charts, graphs, and visual representations of data can enhance the readability and impact of the reports.

Transparency and accuracy are paramount in reporting milestones and achievements. Site managers should avoid glossing over challenges or setbacks and provide honest assessments of project progress. In case of any delays or issues, it is crucial to outline the remedial actions being taken to mitigate risks and get the project back on track.

In addition to written reports, site managers may organize periodic meetings with stakeholders to discuss project updates and address any questions or concerns. These meetings serve as opportunities for face-to-face communication and foster a collaborative atmosphere between the project team and stakeholders.

Flexibility in reporting is essential, as stakeholders may have varying information needs. Some stakeholders may require high-level summaries, while others may require in-depth technical details. Site managers should tailor the reporting approach to cater to different audiences while maintaining the core message and key project updates.

Continuous improvement is crucial in the reporting process. Site managers should regularly seek feedback from stakeholders on the effectiveness of the reports and make necessary adjustments to enhance communication and meet stakeholders' needs more effectively.

Identifying and Addressing Potential Schedule Delays

Identifying and addressing potential schedule delays is a crucial aspect of successful project management for site managers. Delays can arise due to various reasons, such as unforeseen challenges, changes in project scope, weather conditions, material availability, or subcontractor performance. It is essential for site managers to be proactive in detecting potential delays and taking prompt action to mitigate their impact on the project timeline.

One of the key steps in identifying potential schedule delays is to establish a well-defined project schedule at the outset. This schedule should include all the necessary tasks, their dependencies, and the estimated duration for each activity. By having a clear roadmap, site managers can easily track progress and identify any deviations from the original plan.

Regularly monitoring progress against the schedule is vital to detect early warning signs of delays. Site managers should use project management tools and techniques, such as Gantt charts, critical path analysis, and earned value management, to assess the current status and forecast future project milestones. Early detection of delays allows site managers to take proactive measures before they escalate into more significant issues.

When potential delays are identified, site managers should investigate the root causes and assess their impact on the overall project timeline. This may involve collaborating with project teams, subcontractors, and stakeholders to understand the challenges and explore possible solutions.

Implementing corrective actions is essential to address potential delays. Site managers should work with the project team to develop realistic recovery plans, adjust resource allocation, and optimize workflow to bring the project back on track. It is crucial to communicate these adjustments effectively to all stakeholders to ensure a unified understanding of the revised timeline.

Risk management plays a significant role in mitigating schedule delays. Site managers should continuously assess potential risks that could impact the project schedule and have contingency plans in place to respond promptly if these risks materialize.

Maintaining effective communication is vital throughout the process of identifying and addressing schedule delays. Site managers should keep all stakeholders informed about the situation, the actions being taken, and the potential impact on the project timeline. Regular progress updates and transparent communication can help build trust and support from stakeholders.

Effective Progress Reporting to Stakeholders
Effective progress reporting to stakeholders is crucial for keeping all parties informed about the status of a construction project. Site managers play a vital role in ensuring that progress is accurately and transparently communicated to stakeholders. Here are some key strategies for effective progress reporting:

Clear and Concise Updates: Site managers should provide regular updates that are clear, concise, and easy to understand. Avoid technical jargon and use language that is accessible to all stakeholders, including non-technical individuals.

Regular Reporting Schedule: Establish a regular reporting schedule to provide updates at predetermined intervals, such as weekly or monthly. Consistency in reporting helps stakeholders anticipate and plan for the information they will receive.

Key Performance Indicators (KPIs): Identify and track key performance indicators that are relevant to the project's success. These may include milestones achieved, budget performance, safety metrics, and any other critical project metrics.

Visual Representation: Use charts, graphs, and other visual aids to present data in an easily digestible format. Visual representation can help stakeholders quickly grasp project progress and trends.

Highlighting Achievements and Challenges: In addition to reporting on progress, also highlight notable achievements and milestones reached. Similarly, be transparent about any challenges or issues that have arisen and the steps being taken to address them.

Forecasting: Provide stakeholders with a forward-looking outlook by forecasting future milestones and potential risks. This can help stakeholders plan and make informed decisions.

Customized Reporting: Tailor progress reports to meet the specific needs of different stakeholders. For example, investors may be more interested in financial metrics, while project team members may focus on construction progress.

Engage in Two-Way Communication: Progress reporting should not be a one-way communication process. Encourage stakeholders to provide feedback and ask questions. Respond promptly to inquiries and address any concerns raised.

Use Technology: Leverage project management software and other digital tools to streamline progress reporting. Technology can help automate data collection and reporting processes, improving accuracy and efficiency.

Consolidated Reporting: If there are multiple projects or workstreams, consider providing consolidated reports that give stakeholders a holistic view of the overall progress and performance.

Adaptability: Be prepared to adapt the reporting approach based on stakeholder feedback and changing project requirements. Flexibility in reporting ensures that information remains relevant and valuable.

Celebrate Success: Acknowledge and celebrate project successes with stakeholders. Recognizing achievements fosters a positive project culture and encourages continued support.

Providing Regular Project Updates to Clients and Stakeholders
As a site manager, one of your key responsibilities is to provide regular project updates to clients and stakeholders. Effective communication and transparency are vital for ensuring that all relevant parties are well-informed about the project's progress, challenges, and achievements. Regular updates help build trust, foster collaboration, and demonstrate your commitment to the project's success.

At the outset of the project, develop a comprehensive communication plan that outlines the frequency, method, and content of project updates. This plan should consider the specific needs and preferences of clients and stakeholders. Decide on the appropriate frequency for project updates. This could be weekly, bi-weekly, or monthly, depending on the project's timeline and complexity. Consistency in updates is crucial for keeping stakeholders engaged and informed.

Understand the varying levels of technical expertise among clients and stakeholders. Tailor the updates to the needs of the audience, avoiding jargon and technical language that may be difficult for non-experts to understand. Include key project metrics and performance indicators in the updates. Highlight progress against the project timeline, budget status, safety records, and other relevant KPIs. Providing data-driven insights helps stakeholders assess the project's health.

Use charts, graphs, and visuals to present data in a clear and concise manner. Visual aids enhance understanding and make it easier for stakeholders to grasp complex information quickly. Acknowledge and celebrate project milestones and achievements in the updates. Recognizing success not only boosts team morale but also demonstrates progress to clients and stakeholders.

Be transparent about any challenges or risks the project is facing. Clearly communicate the steps being taken to address these issues and minimize their impact on the project's objectives. Foster an open and collaborative environment by encouraging feedback and questions from clients and stakeholders. Respond to inquiries promptly and provide clarifications as needed.

Employ a mix of communication channels to provide updates. In addition to written reports, consider using emails, phone calls, virtual meetings, or in-person presentations to ensure stakeholders are well-informed. Be prepared to adapt your communication approach as the project progresses. Stakeholders' needs may change over time, and flexibility in your updates is essential to meet their evolving requirements.

Respect stakeholders' time by keeping updates concise and focused on the most relevant information. Avoid overwhelming them with unnecessary details. Maintain a record of all project updates and relevant documentation for future reference. Proper archiving ensures a clear historical record of the project's progress.

Communicating Progress, Risks, and Mitigation Plans

As a site manager, one of your primary responsibilities is to effectively communicate the project's progress, risks, and mitigation plans to clients and stakeholders. Open and transparent communication is essential for building trust, managing expectations, and ensuring the successful execution of the project.

When communicating progress, provide a clear and concise overview of the project's status. Highlight key achievements, milestones reached, and any significant developments since the last update. Use data and metrics to support your updates, such as percentage completion, timelines, and budgetary information. Visual aids like charts and graphs can enhance understanding and make complex information more accessible.

In addition to progress updates, be proactive in communicating any potential risks that may impact the project's timeline, budget, or quality. Identify and assess these risks, and develop robust mitigation plans to address them. Clearly outline the steps being taken to minimize or eliminate risks and ensure the project stays on track. Being transparent about risks and mitigation plans demonstrates your commitment to managing challenges effectively.

Adapt your communication style to suit the needs and preferences of different stakeholders. Some may require detailed technical information, while others may prefer high-level summaries. Tailor your updates accordingly to ensure that all stakeholders can easily grasp the information presented.

Maintain a regular communication schedule, and be consistent in providing updates. Depending on the project's timeline and complexity, consider holding weekly or bi-weekly progress meetings with stakeholders. Use these opportunities to discuss progress, address concerns, and gather feedback from stakeholders.

When communicating risks and mitigation plans, focus on practical and actionable steps. Avoid using technical jargon and explain concepts in a clear and straightforward manner. Encourage stakeholders to ask questions and provide feedback, and be responsive to their inquiries.

Use various communication channels to reach stakeholders effectively. In addition to formal meetings, utilize emails, reports, and project management software to share updates and documents. Ensure that all stakeholders have access to the necessary information and documentation.

As the project evolves, be prepared to adjust your communication approach. Stakeholder needs and priorities may change, and it is essential to adapt your updates to address their evolving concerns.

Addressing Stakeholder Concerns and Requests

As a site manager, one of your primary responsibilities is to manage and address the concerns and requests of various stakeholders involved in the construction project. Stakeholders may include clients, contractors, suppliers, regulatory authorities, and community members. Effectively handling their concerns and requests is essential for maintaining positive relationships, ensuring project success, and fostering a collaborative and productive work environment.

Active Listening and Understanding Stakeholder Needs

The first step in addressing stakeholder concerns and requests is to practice active listening. When stakeholders express their issues or make requests, give them your full attention and take the time to understand their needs thoroughly. Ask questions to clarify any ambiguities and gather all relevant information. By actively listening, you demonstrate empathy and respect for their perspective, which can go a long way in building trust.

Timely and Transparent Communication

Timely communication is vital when dealing with stakeholder concerns and requests. Acknowledge their messages promptly and provide an estimated timeframe for addressing the issues. If more time is needed for a comprehensive response, communicate that too. Transparent communication is key to managing expectations and maintaining credibility with stakeholders.

Thorough Investigation and Analysis

Once you have a clear understanding of the concerns or requests, conduct a thorough investigation. Gather all pertinent data and involve relevant team members if necessary. Analyze the information to determine the root cause of the problem and identify potential solutions or courses of action.

Offering Practical Solutions

When addressing stakeholder concerns, it is crucial to propose practical and feasible solutions. Consider the project's constraints, budget, and schedule while exploring options. Present stakeholders with well-considered alternatives, and if necessary, collaborate with them to find the best way forward.

Collaboration and Negotiation

In some cases, stakeholder concerns or requests may involve trade-offs or require negotiations. Be open to collaboration and negotiation to find mutually beneficial solutions. Engage in constructive discussions to balance stakeholders' needs with project requirements.

Documenting and Following Up

Keep detailed records of all stakeholder interactions, concerns, and requests, along with the actions taken to address them. Documentation ensures accountability and helps in monitoring the progress of resolving issues. After implementing solutions, follow up with stakeholders to ensure that they are satisfied with the outcome.

Proactive Approach

Adopt a proactive approach to stakeholder management by anticipating potential concerns and addressing them before they escalate. Regularly provide updates on project progress and address any potential risks or issues that stakeholders may be concerned about.

Continuous Improvement

Use feedback from stakeholders as an opportunity for continuous improvement. Learn from their input and experiences to enhance communication and project management processes. Regularly seek feedback to ensure that stakeholder needs are consistently met.

Mitigating Delays and Managing Project Changes

In the construction industry, delays and changes are almost inevitable due to various factors such as unforeseen site conditions, weather disruptions, material availability, design modifications, or scope adjustments. As a site manager, it is crucial to proactively identify potential delays and changes and implement strategies to mitigate their impact on the project's timeline and budget.

Comprehensive Planning and Risk Assessment

At the outset of the project, conduct a thorough planning process that includes a comprehensive risk assessment. Identify potential risks and uncertainties that could lead to delays or changes and develop contingency plans to address them. By anticipating challenges in advance, you can be better prepared to respond when they arise.

Regular Project Monitoring and Reporting

Implement a robust project monitoring and reporting system to keep track of project progress and identify any early signs of delays or changes. Regularly review project metrics, such as schedule adherence, productivity rates, and resource utilization, to detect deviations from the plan.

Collaborative Decision-Making

Involve all relevant stakeholders, including clients, contractors, and subcontractors, in the decision-making process. Collaboration fosters a sense of ownership and accountability among the team and helps in finding innovative solutions to mitigate delays and manage changes effectively.

Contingency Planning and Flexibility

Incorporate contingency plans into the project schedule to account for potential delays or changes. Allow for flexibility in the timeline and resource allocation to accommodate unforeseen circumstances without jeopardizing the project's overall objectives.

Effective Communication

Maintain clear and open lines of communication with all stakeholders. Timely communication of any changes or delays is crucial to manage expectations and minimize disruptions. Address concerns promptly and provide regular updates to keep everyone informed of the project's status.

Change Management Process
Establish a formal change management process to assess and approve any requested changes to the project scope, schedule, or budget. This process should include a thorough evaluation of the impact of the proposed change and its alignment with project goals.

Resource Optimization
Optimize the allocation of resources to prevent bottlenecks and potential delays. Ensure that skilled labor, materials, and equipment are available when needed and that resources are utilized efficiently throughout the project lifecycle.

Lean Construction Practices
Implement lean construction practices that focus on eliminating waste, streamlining processes, and enhancing productivity. Through lean principles, you can reduce the likelihood of delays and improve overall project efficiency.

Analyzing Root Causes
When delays or changes occur, conduct a root cause analysis to identify the underlying reasons. Learning from past experiences enables you to implement preventive measures and avoid similar issues in future projects.

Documenting Changes and Lessons Learned
Maintain accurate documentation of all changes, delays, and their resolutions. Use this documentation as a resource for future projects and to apply lessons learned in improving project management practices.

Mitigating delays and managing project changes requires a proactive and collaborative approach. By being prepared, flexible, and communicative, site managers can effectively navigate through challenges and ensure successful project outcomes. The ability to address delays and changes with resilience and adaptability is a hallmark of an effective site manager in the construction industry.

Identifying Causes of Delays and Disruptions

Identifying the causes of delays and disruptions is a crucial responsibility for site managers to ensure the smooth execution of construction projects. Delays and disruptions can have significant impacts on the project's timeline, budget, and overall success. By proactively identifying the root causes of these issues, site managers can implement effective strategies to mitigate their effects and keep the project on track.

One of the key aspects of identifying causes is conducting a thorough analysis of the project's progress and performance. This involves reviewing the project schedule and comparing it to the actual progress made at various stages. It also entails assessing the productivity of the workforce, availability and delivery of materials, and the performance of subcontractors and suppliers. By closely monitoring these factors, site managers can pinpoint areas where delays and disruptions may be occurring.

Additionally, site managers should regularly communicate with the project team, including contractors, subcontractors, and suppliers, to gain insights into any challenges they may be facing. Collaborative discussions can provide valuable information on potential causes of delays, such as design changes, weather-related issues, or equipment malfunctions. Engaging with the team fosters a culture of open communication, allowing for the timely identification and resolution of problems.

Another essential aspect is reviewing the project's documentation, such as construction drawings, specifications, and contracts. Discrepancies or ambiguities in these documents can lead to misunderstandings and delays. A meticulous examination of project documentation can help identify any potential areas of concern that may need clarification or revision.

Environmental factors, such as adverse weather conditions, geological challenges, or regulatory hurdles, can also contribute to delays and disruptions. Site managers must be aware of these external influences and take them into account when planning and executing the project.

Furthermore, site managers should assess their own management practices and decision-making processes. Inefficient resource allocation, poor coordination, or lack of effective communication within the team can all be factors contributing to delays and disruptions. By introspecting and evaluating their own management approach, site managers can implement improvements and streamline operations.

Lastly, historical data and lessons learned from previous projects can offer valuable insights into the causes of delays and disruptions. Analyzing past experiences allows site managers to anticipate potential challenges and implement preventive measures in future projects.

Identifying the causes of delays and disruptions requires a comprehensive and systematic approach. Site managers should employ data-driven analysis, open communication with the team, thorough document review, consideration of external factors, and self-assessment to gain a holistic understanding of the project's performance. By doing so, site managers can proactively address challenges and ensure the successful execution of construction projects.

Implementing Strategies for Schedule Recovery

Implementing strategies for schedule recovery is a critical aspect of effective construction project management. When delays occur, it is essential for site managers to take swift and decisive action to bring the project back on track and meet the established deadlines. The goal is to optimize the project schedule and ensure that any lost time is regained without compromising the quality of work.

One of the first steps in schedule recovery is to conduct a comprehensive analysis of the causes of the delays. By understanding the root causes, site managers can develop targeted strategies to address each issue effectively. This may involve revising the project schedule, reallocating resources, or renegotiating contracts with subcontractors and suppliers.

A common approach to schedule recovery is the adoption of fast-tracking and crashing techniques. Fast-tracking involves overlapping certain project activities that were initially planned to be performed sequentially. By doing so, critical tasks can be completed simultaneously, reducing the overall project duration. However, fast-tracking may increase the risk of potential conflicts and requires careful coordination among the project team.

On the other hand, crashing involves injecting additional resources into critical activities to expedite their completion. This may include adding more workers, increasing equipment usage, or working overtime. While crashing can accelerate progress, it may also lead to additional costs, and site managers must weigh the benefits against the potential drawbacks.

Resource reallocation is another valuable strategy for schedule recovery. By redistributing resources from non-critical activities to those that are behind schedule, site managers can focus on completing the most crucial tasks promptly. This requires a careful evaluation of the project's priorities and constant monitoring to ensure optimal resource utilization.

Furthermore, collaborative problem-solving with the project team is essential. By involving all stakeholders, including contractors, subcontractors, suppliers, and designers, in the recovery process, site managers can gain valuable insights and innovative ideas for expediting progress.

Site managers should also consider the use of technology and software to streamline project management and facilitate schedule recovery. Advanced project management tools can help in real-time monitoring, resource allocation, and communication, enabling faster decision-making and greater efficiency.

Another aspect of schedule recovery is setting realistic and achievable milestones. By breaking down the project into manageable phases and regularly reviewing progress, site managers can detect potential delays early and take appropriate corrective measures.

Lastly, continuous communication with the project owner and other stakeholders is crucial. Transparent reporting of progress, challenges, and recovery efforts instills confidence and trust, ensuring everyone is aware of the situation and committed to achieving the project's goals.

Implementing strategies for schedule recovery requires a proactive and multifaceted approach. Site managers should analyze the causes of delays, employ fast-tracking and crashing techniques judiciously, reallocate resources strategically, collaborate with the project team, leverage technology, set realistic milestones, and maintain open communication with stakeholders. By doing so, site managers can effectively recover the schedule, keep the project on track, and deliver successful construction projects.

Managing Change Orders and Variations

Managing change orders and variations is an integral part of construction project management. Change orders refer to modifications or alterations to the original scope of work, while variations pertain to deviations from the initial contract specifications. These changes can arise due to various factors, such as design revisions, unforeseen site conditions, client preferences, or regulatory requirements.

To effectively manage change orders and variations, site managers must adopt a systematic approach that ensures transparency, fairness, and adherence to contractual obligations. Here are essential steps and considerations in the process:

Documentation and Record-Keeping: Maintaining accurate and comprehensive records is paramount. Site managers should document all changes, including the reasons for the variations, their impact on the project schedule, costs, and any agreed-upon adjustments to the scope of work. Proper documentation serves as evidence and provides a basis for negotiations and dispute resolution if required.

Evaluation of Change Requests: When a change request is received, site managers should carefully assess its feasibility, impact on the project timeline, and cost implications. Collaborating with the project team, including architects, engineers, and subcontractors, is crucial to understand the technical and logistical aspects of the proposed changes.

Communication with Stakeholders: Transparent communication is vital in managing change orders and variations. Site managers should promptly inform the client, subcontractors, and other relevant parties about the proposed changes, their consequences, and the potential impact on the project's budget and timeline. Regular updates on the status of change orders are essential to maintain trust and mitigate misunderstandings.

Negotiation and Agreement: Negotiating fair and reasonable terms for change orders is essential to ensure a win-win outcome for all parties involved. Site managers should engage in constructive discussions with clients and subcontractors to reach mutually acceptable terms and avoid disputes. The negotiated terms should be documented in written agreements to avoid ambiguity.

Impact on Project Schedule: Change orders and variations can have significant implications for the project timeline. Site managers should evaluate the effects of changes on critical activities and milestones. If necessary, a revised project schedule should be developed to accommodate the alterations while minimizing disruptions.

Cost Management: Change orders often involve cost adjustments. Site managers should diligently estimate the additional costs associated with variations and ensure that pricing is fair and reasonable. Cost control measures should be implemented to prevent cost overruns and ensure that project budgets remain within acceptable limits.

Contractual Compliance: Throughout the process of managing change orders and variations, site managers must adhere to the terms and conditions outlined in the contract. This includes ensuring that all contractual requirements for change orders are met, and proper approvals are obtained from the client and other relevant parties.

Dispute Resolution: Despite best efforts, disputes related to change orders may arise. Site managers should be prepared to handle such situations by resorting to established dispute resolution mechanisms, such as mediation or arbitration, as specified in the contract.

Managing change orders and variations is a complex and critical aspect of construction project management. Site managers should focus on documentation, evaluation, transparent communication, fair negotiation, impact assessment on the project schedule, cost management, contractual compliance, and dispute resolution. By adopting a methodical and collaborative approach, site managers can effectively navigate through change orders and variations, ensuring successful project delivery and client satisfaction.

CHAPTER 9:
BUDGETING AND
COST CONTROL

B udgeting and cost control are essential for construction project management. A well-planned budget sets the financial framework, while cost control prevents overruns. Site managers allocate resources wisely, track expenses, implement cost-saving measures, and address deviations promptly. Effective budgeting ensures project profitability and success.

Estimating and Budgeting
for On-Site Operations

Estimating and budgeting play a crucial role in the successful execution of construction projects. As a site manager, your ability to create accurate estimates and well-structured budgets is paramount to ensuring that the project stays on track financially and meets its objectives. The process of estimating and budgeting involves a comprehensive evaluation of all the resources required for the project, including labor, materials, equipment, and other associated costs.

To begin, thorough planning and analysis are necessary. This involves carefully examining the project scope, architectural drawings, and specifications. Collaborating with the project team, including architects, engineers, and subcontractors, will help in understanding the intricacies and complexities of the project. Additionally, historical data from past projects can be valuable in providing insights into similar tasks, costs, and challenges.

Accurate estimation is a blend of science and experience. Site managers need to account for various factors, including market prices for materials and labor, inflation, and potential risks that may arise during the project's lifecycle. Using industry-standard estimating software can streamline the process and enhance accuracy.

Once the estimates are in place, the next step is to develop a comprehensive budget. The budget serves as a financial blueprint for the project, detailing all anticipated expenses and income streams. Effective budgeting requires meticulous attention to detail and a forward-thinking approach to anticipate any potential financial challenges.

As site managers, you must be proactive in managing the budget throughout the project's lifecycle. Regularly monitor and track expenses against the budgeted amounts to identify any deviations early on. Implementing a robust cost control system allows for real-time analysis of financial performance, enabling you to make informed decisions and take corrective actions promptly.

It is essential to communicate the budget to all stakeholders, including the project owner, investors, and team members. Transparency in financial matters fosters trust and demonstrates your commitment to delivering the project within the agreed-upon financial parameters.

Furthermore, budgeting for on-site operations should not be a one-time exercise. As the project progresses, continuous evaluation and adjustment may be necessary to accommodate changes, unforeseen circumstances, and evolving requirements. By being adaptable and responsive to such fluctuations, you can maintain the financial health of the project and ensure its successful completion.

Estimating and budgeting for on-site operations are critical skills that site managers must master. Accurate estimates and well-structured budgets provide the foundation for a successful project and help mitigate financial risks. By implementing effective cost control measures and proactive financial management, site managers can navigate the complexities of construction projects with confidence and achieve successful outcomes.

Conducting Accurate Project Cost Estimation
Project cost estimation is a fundamental aspect of construction management that requires precision and meticulous attention to detail. As a site manager, your ability to conduct accurate cost estimation is instrumental in ensuring the success of a construction project. An accurate cost estimation sets the groundwork for effective budgeting, resource allocation, and financial planning.

The first step in conducting project cost estimation is to thoroughly understand the project scope and requirements. Reviewing architectural plans, engineering specifications, and any other relevant documents will provide valuable insights into the complexity and scale of the project. Collaborating with subject matter experts, such as architects, engineers, and experienced contractors, can also contribute to a more comprehensive understanding of the project's nuances and potential cost drivers.

To arrive at accurate cost estimates, it is essential to break down the project into its individual components and activities. This process, known as a work breakdown structure (WBS), allows for a detailed analysis of the labor, materials, equipment, and other resources required for each task. Additionally, past project data and historical cost records can serve as valuable references to identify patterns, trends, and benchmarks for similar projects.

A significant aspect of project cost estimation is accurately quantifying the quantities of materials and resources needed. This includes calculating the precise amount of construction materials, labor hours required, and equipment usage. Taking into account factors such as market prices, inflation, and regional variations will contribute to more realistic estimates.

Leveraging specialized construction estimating software can greatly streamline the cost estimation process. These tools help site managers organize data efficiently, generate detailed cost breakdowns, and perform complex calculations with increased accuracy. Moreover, the software often allows for real-time updates, enabling site managers to adjust estimates promptly as project parameters change.

As a site manager, you must also be vigilant in accounting for potential risks and uncertainties that may impact the project's cost. Contingency allowances should be included in the estimation to address unforeseen events and mitigate potential cost overruns.

Effective communication with stakeholders is vital throughout the cost estimation process. Clearly presenting the cost estimates, assumptions, and methodologies to project owners, investors, and other team members fosters transparency and builds confidence in the project's financial viability.

Developing Comprehensive Budget Plans

Creating a comprehensive budget plan is a crucial task for site managers to ensure the successful execution of construction projects. A well-developed budget plan serves as a financial roadmap, guiding the allocation of resources, controlling costs, and achieving project objectives. It requires a methodical approach, attention to detail, and the ability to adapt to changing circumstances.

The first step in developing a budget plan is to gather all relevant cost estimates obtained during the project's estimation phase. These estimates should encompass all aspects of the construction project, including labor, materials, equipment, subcontractor services, permits, and other expenses. Site managers must meticulously review and analyze these estimates to ensure accuracy and completeness.

Once the cost estimates are compiled, the next step is to organize them into a structured budget framework. This framework should break down the costs into specific categories and align with the work breakdown structure (WBS) of the project. By doing so, site managers can allocate the budget to individual tasks, phases, or milestones, facilitating better financial management and control.

A comprehensive budget plan must also account for contingencies and potential risks. In construction projects, uncertainties are common, and unexpected events can impact the budget. Allocating a contingency reserve in the budget plan ensures that there are funds available to address unforeseen circumstances without derailing the project's progress.

A critical aspect of budget planning is maintaining a balance between cost optimization and quality delivery. Site managers must work closely with project teams and stakeholders to identify areas where cost savings can be achieved without compromising the project's integrity, safety, or performance.

Moreover, the budget plan should align with the project schedule and timeline. Integrating the budget and schedule ensures that financial resources are available when needed, avoiding delays caused by financial constraints. Regularly tracking budget performance against actual expenditures enables site managers to identify any deviations and implement corrective actions promptly.

In addition to internal stakeholders, site managers must collaborate with the finance department or project accounting team to ensure the budget plan adheres to organizational financial guidelines and reporting standards. Proper documentation and transparency in financial reporting enhance accountability and build trust among stakeholders.

Throughout the project's lifecycle, site managers must continuously monitor the budget plan and be prepared to adapt to changing circumstances. This adaptability may involve revising the budget to accommodate scope changes, cost variations, or unexpected challenges that arise during construction.

Factoring in Contingencies and Risk Management
In the dynamic world of construction, uncertainties and risks are inherent. Aspiring and current site managers must recognize the importance of factoring in contingencies and implementing effective risk management strategies to ensure successful project delivery. The ability to anticipate potential challenges and develop appropriate response plans is a key aspect of responsible project management.

Contingencies refer to provisions made in the budget and schedule to account for unforeseen events that may impact the project's progress. These events could include changes in scope, material price fluctuations, weather disruptions, labor issues, or delays caused by external factors. By allocating a contingency reserve, site managers create a financial buffer that allows them to respond swiftly to unexpected circumstances without compromising the project's objectives.

Risk management is a proactive approach to identify, assess, and mitigate potential risks before they escalate into major issues. Site managers must conduct a thorough risk assessment at the project's outset, analyzing various aspects such as technical, environmental, financial, and contractual risks. This assessment helps in understanding the probability and potential impact of each risk, thereby enabling the development of appropriate response strategies.

Effective risk management involves a structured approach

Risk Identification: Site managers, in collaboration with project teams, must identify all potential risks that could affect the project. Brainstorming sessions, historical data analysis, and input from stakeholders are valuable tools to identify risks comprehensively.

Risk Assessment: Once identified, each risk must be evaluated based on its likelihood of occurrence and potential consequences. By prioritizing risks based on severity, site managers can focus their efforts on addressing the most critical ones.

Risk Mitigation: After assessing the risks, site managers should develop mitigation plans to reduce the likelihood and impact of adverse events. These plans may involve process improvements, alternative sourcing strategies, or implementing specific safety measures.

Risk Monitoring: Throughout the project's duration, site managers must continuously monitor identified risks and assess their evolution. Timely identification of changes in risk levels allows for proactive responses to emerging threats.

Contingency Execution: In case a risk materializes, the contingency plan comes into action. Site managers must swiftly deploy the appropriate response measures to minimize the impact on the project's schedule, budget, and overall performance.

Learning from Risks: At the conclusion of the project, site managers should conduct a comprehensive review of risk management processes. This review aims to identify lessons learned and areas of improvement for future projects.

By integrating contingencies and risk management into the project's planning and execution, site managers demonstrate their commitment to proactive and responsible project management. This approach not only enhances the project's chances of success but also instills confidence in clients, stakeholders, and project teams by showcasing a readiness to address challenges with efficiency and competence.

Monitoring Costs and Controlling Expenditures

Cost management is a crucial aspect of successful project execution, and aspiring and current site managers must adopt a vigilant approach to monitor costs and control expenditures effectively. In a construction project, expenses can quickly escalate, impacting the project's budget and overall profitability. Hence, site managers need to implement robust cost monitoring and control mechanisms to ensure that the project stays on track financially.

Budget Adherence: Site managers must start by developing a comprehensive budget that covers all aspects of the project, including labor, materials, equipment, subcontractors, and contingencies. Adhering to the approved budget becomes the foundation of cost control.

Expense Tracking: Monitoring project costs requires a diligent tracking system. Site managers should record all expenses accurately and regularly update the budget to reflect the actual financial status of the project.

Variance Analysis: Regularly comparing actual costs to the budgeted costs enables site managers to identify any variances. Variance analysis allows them to pinpoint areas where expenditures may be exceeding the allocated budget.

Identifying Cost Overruns: When cost overruns occur, site managers must take immediate action to understand the underlying causes. By identifying the root causes of overruns, they can implement corrective measures promptly.

Change Order Management: Changes in project scope often lead to additional costs. Site managers should meticulously manage change orders and ensure that they are appropriately approved, documented, and incorporated into the budget.

Vendor and Subcontractor Negotiations: Effective negotiation skills play a vital role in controlling expenditures. Site managers should negotiate competitive prices with vendors and subcontractors without compromising on the quality of goods and services.

Value Engineering: Employing value engineering practices can help optimize project costs without sacrificing performance or quality. It involves evaluating alternatives and selecting the most cost-effective solutions.

Cost Forecasting: As the project progresses, site managers should utilize historical cost data and projections to forecast future expenses. This enables them to anticipate potential cost fluctuations and take preemptive actions.

Controlling Non-Essential Costs: Site managers should identify non-essential costs that do not contribute significantly to the project's success. Reducing or eliminating such costs can free up resources for critical project components.

Regular Reporting: Timely and transparent cost reporting to stakeholders, including clients and project sponsors, fosters accountability and trust. Clear communication about the project's financial health enables stakeholders to make informed decisions.

Auditing and Compliance: Regular financial audits and

compliance checks help ensure that the project adheres to financial regulations and contractual obligations.

Effective cost monitoring and expenditure control are vital to the project's success and the site manager's reputation. By implementing these practices, site managers can maintain financial discipline, deliver projects within budget, and uphold the organization's credibility in the construction industry.

Tracking Project Expenses and Budget Variances

Aspiring and current site managers must diligently track project expenses and closely monitor budget variances to ensure the financial success of construction projects. The ability to track and manage project costs is fundamental to completing projects within budget and delivering value to stakeholders. Here are essential practices for tracking expenses and managing budget variances effectively:

Comprehensive Expense Tracking: Site managers should establish a robust system for recording all project expenses, including labor, materials, equipment, subcontractors, permits, and miscellaneous costs. Accurate and comprehensive expense tracking forms the foundation for successful budget management.

Real-Time Data Collection: Utilizing modern technology and software solutions can help site managers collect and update expense data in real-time. This enables them to make informed decisions promptly, mitigating potential financial risks.

Budget vs. Actual Analysis: Regularly comparing actual expenses to the budgeted amounts allows site managers to identify budget variances. This analysis enables them to understand the deviations and take corrective actions when necessary.

Early Identification of Variances: Site managers should be proactive in identifying budget variances as soon as they occur.

Early detection allows them to implement timely corrective measures and prevent further deviations.

Communication with Stakeholders: Transparent communication about budget variances is essential to maintain trust and credibility with stakeholders. Site managers should regularly update project sponsors and clients on any financial challenges and the steps being taken to address them.

Variance Root-Cause Analysis: Investigating the root causes of budget variances is critical to finding effective solutions. Whether caused by unforeseen changes, scope creep, or material price fluctuations, understanding the reasons behind variances helps in formulating appropriate responses.

Contingency Utilization: Properly managing contingency reserves is crucial for addressing unexpected expenses without affecting the primary budget. Site managers should carefully assess when to utilize contingency funds and ensure that it aligns with the project's priorities.

Change Order Management: Changes in project scope often lead to budget adjustments. Site managers must diligently manage change orders, ensuring they are documented, reviewed, and approved by relevant stakeholders.

Forecasting and Projection: Utilizing historical data and project trends, site managers can forecast future expenses and potential budget variances. Accurate projections facilitate better financial planning and resource allocation.

Budget Revisions: When faced with significant budget variances, site managers may need to revise the budget and align it with the project's current realities. This ensures that financial decisions are based on the most up-to-date information available.

Compliance and Audit: Regular audits and compliance checks contribute to financial discipline and ensure that the project adheres to contractual obligations and financial regulations.

Effective expense tracking and budget variance management are essential skills for site managers. Through these practices and leveraging technology for data analysis, site managers can maintain control over project finances, deliver successful projects, and build a reputation for financial stewardship in the construction industry.

Implementing Cost Control Measures and Forecasting
Cost control is a critical aspect of effective project management, and site managers play a pivotal role in ensuring projects stay within budgetary constraints. By implementing cost control measures and accurate forecasting techniques, site managers can proactively manage expenses and make informed decisions to achieve financial success.

At the outset of a project, site managers should allocate budget resources to various project components, including labor, materials, equipment, and contingency funds. This step ensures that financial resources are appropriately distributed and aligned with project priorities.

To track project expenses accurately, site managers should establish robust systems, using modern project management software and tools that can streamline data collection. Real-time expense monitoring enables better decision-making and allows early identification of potential cost overruns or deviations.

Including contingency funds in the budget is crucial to address unforeseen events and risks. Site managers must proactively manage contingency reserves to prevent unexpected expenses from affecting the primary budget.

Managing change orders effectively is vital for controlling project costs. Site managers should carefully evaluate change requests, assess their impact on the budget, and obtain proper approvals before implementation.

Encouraging value engineering practices helps optimize project costs without compromising quality. By seeking cost-effective alternatives and innovative solutions, site managers can achieve cost savings without sacrificing project objectives.

Skilled negotiation with vendors and subcontractors can lead to favorable pricing and terms. Site managers should aim to secure competitive rates and favorable payment conditions to optimize cost control.

Utilizing data analytics and historical project data, site managers can employ forecasting techniques to anticipate future expenses and potential budget variances. Accurate forecasting enables proactive planning and risk management.

Identifying potential risks that may impact project costs is essential. Site managers should conduct comprehensive risk assessments and develop mitigation strategies to address these risks proactively.

Implementing Earned Value Management (EVM) methodologies enables site managers to measure project performance against the budget and schedule. EVM provides valuable insights into cost and schedule variances, aiding in better decision-making.

Transparent and regular reporting to project stakeholders helps keep everyone informed about project financials. Site managers should provide clear and concise reports, highlighting budget status, cost-saving measures, and any deviations from the plan.

Through these cost control measures and implementing forecasting techniques, site managers can effectively manage project expenses, optimize resource allocation, and deliver successful projects within budgetary constraints. The ability to control costs and forecast accurately contributes significantly to the success of construction projects and enhances a site manager's reputation as a skilled and proficient project leader.

Identifying Cost Reduction Opportunities

As a site manager, your role involves identifying and capitalizing on cost reduction opportunities throughout the project. Efficient cost management not only contributes to the project's financial success but also enhances overall efficiency and competitiveness. To achieve this, consider implementing the following strategies:

Begin by conducting a comprehensive cost analysis of the entire project. Break down the budget into various components such as labor, materials, equipment, and overhead expenses. This analysis will provide a clear picture of where the majority of expenses lie and where potential cost-saving opportunities may exist.

Value engineering is a valuable approach to explore cost-saving alternatives without compromising on quality. Collaborate with the design team, engineers, and architects to scrutinize the project's design and specifications, aiming to achieve the same functionality at a lower cost.

Review your existing vendor and supplier contracts to ensure that you are receiving the best possible prices and terms. Consider seeking competitive bids from multiple vendors to identify cost-saving options while maintaining quality standards.

Consolidate purchases of materials and equipment to take advantage of bulk pricing. Negotiate with suppliers to secure discounts, especially for long-term or high-volume purchases.

Monitor labor and equipment utilization regularly to ensure that resources are efficiently allocated. Avoid overstaffing or underutilization of equipment, which can lead to unnecessary costs.

Implement waste management practices to minimize material wastage on the construction site. Encourage recycling and reuse of materials whenever possible.

Embrace energy-efficient construction techniques and technology to reduce utility expenses during the project's operational phase.

Regularly maintain and service equipment to prevent breakdowns and costly repairs. Planned maintenance can extend the lifespan of machinery and reduce downtime.

Conduct risk assessments to identify potential cost-incurring events and develop contingency plans to mitigate their impact.

Foster a culture of continuous improvement within the project team. Encourage open communication and feedback to identify cost-saving ideas from all stakeholders.

Benchmark your project's performance against similar past projects to identify areas where cost reduction can be achieved. Analyze historical data and lessons learned to inform future decision-making.

Implement lean construction principles, which aim to eliminate waste and streamline processes. This approach can lead to substantial cost savings and improved project delivery.
By adopting these strategies, site managers can successfully identify cost reduction opportunities and contribute to the overall financial success of the construction project. Proactive cost management not only enhances the project's profitability but also reflects the site manager's skills in optimizing resources and delivering high-quality results.

Cost Optimization and Value Engineering Strategies
As a site manager, one of your primary responsibilities is to ensure that construction projects are executed efficiently and within budget. Cost optimization and value engineering are two critical strategies that can help achieve this goal. Let's explore these strategies in detail:

Cost Optimization

Cost optimization involves managing expenses throughout the project lifecycle without compromising on quality or safety. It starts with a detailed cost analysis, breaking down the budget into various components such as labor, materials, equipment, and overhead expenses. By understanding the project's financial structure, site managers can identify potential cost-saving opportunities and areas where resources can be allocated more efficiently.

One essential aspect of cost optimization is value for money. It's not just about finding the cheapest options but rather seeking the best value for the investment. This means evaluating different alternatives and selecting those that offer the most benefits while keeping costs in check.

Another aspect of cost optimization is effective procurement and contract management. Negotiating favorable terms with suppliers, consolidating purchases, and exploring discounts can all contribute to cost reduction. Regularly reviewing vendor contracts and exploring competitive bids are essential practices.

Value Engineering

Value engineering is a systematic and creative approach to improving the project's value by optimizing costs. It involves collaboration between the project team, including engineers, architects, and contractors, to analyze the design and specifications. The goal is to identify alternatives that can provide the same functionality and performance at a lower cost.

Site managers can encourage value engineering workshops where team members brainstorm ideas to achieve cost efficiencies without compromising quality. Value engineering can lead to innovative solutions, more efficient processes, and improved project performance.

Value engineering is not just about cost-cutting; it's about finding ways to enhance the overall value of the project for all stakeholders. This could include improved functionality, increased durability, reduced maintenance costs, or enhanced aesthetics.

Key Considerations

Both cost optimization and value engineering require a proactive and collaborative approach. Site managers should foster a culture of cost consciousness and continuous improvement within the project team. Regular communication and feedback among team members are essential to identify cost-saving opportunities and implement value engineering ideas.

Furthermore, historical data and lessons learned from past projects can provide valuable insights for future cost optimization efforts. Benchmarking against similar projects in the industry can also help identify best practices and areas for improvement.

Cost optimization and value engineering are vital strategies for site managers to effectively manage project budgets while delivering high-quality results. By implementing these strategies, site managers can demonstrate their expertise in resource optimization and contribute significantly to the success of construction projects.

Applying Value Engineering Principles to Maximize Value

As a site manager, one of your core responsibilities is to maximize the value of construction projects for your clients and stakeholders. Value engineering is a powerful tool that can help you achieve this objective. It is a systematic approach that focuses on optimizing the value of a project by identifying innovative solutions and cost-effective alternatives.

Value engineering starts with a thorough understanding of the project's requirements, objectives, and constraints. It involves analyzing the design, materials, and processes to find opportunities for improvement. By engaging the entire project team, including architects, engineers, contractors, and suppliers, you can leverage their expertise to explore different perspectives and ideas.

The key principles of value engineering include

Function Analysis: Value engineering begins with a detailed analysis of the project's functions and objectives. By understanding the core functions that the project must fulfill, you can identify opportunities to enhance performance and efficiency.

Creativity and Brainstorming: Value engineering workshops and brainstorming sessions are essential for generating innovative ideas. Encourage your team members to think outside the box and explore unconventional solutions that can add value to the project.

Cost-Benefit Analysis: Evaluating the cost and benefits of different alternatives is at the heart of value engineering. This involves quantifying the potential savings and improvements resulting from proposed changes.

Life-Cycle Costing: Consider the entire life cycle of the project when evaluating alternatives. This includes not only the initial construction cost but also ongoing maintenance, operation, and replacement costs.

Risk Assessment: Identify and evaluate potential risks associated with proposed changes. While value engineering aims to optimize value, it is crucial to consider potential risks and ensure that proposed solutions do not compromise safety or reliability.

Collaboration and Communication: Effective value engineering requires open communication and collaboration among all stakeholders. Foster a culture of teamwork and encourage active participation from all team members.

Documentation: Proper documentation of value engineering efforts is essential for future reference and decision-making. Keep a record of proposed ideas, evaluations, and decisions made during the value engineering process.

Implementing value engineering principles can lead to a range of benefits, including cost savings, improved performance, reduced environmental impact, and enhanced client satisfaction. It allows site managers to showcase their expertise in delivering projects that are not only on time and within budget but also provide enhanced value to clients and end-users.

To apply value engineering successfully, site managers must stay updated with the latest construction technologies, industry trends, and best practices. Regularly review past projects to identify lessons learned and potential areas for improvement. By integrating value engineering into project management processes, site managers can play a significant role in optimizing value and driving success in construction projects.

Seeking Cost-Efficient Alternatives and Solutions

As a site manager, one of your primary responsibilities is to ensure that construction projects are completed within budget while maintaining the desired level of quality. Seeking cost-efficient alternatives and solutions is a crucial aspect of achieving this goal. By proactively exploring various options, site managers can identify opportunities for cost savings without compromising on the project's overall objectives.

Value Engineering: Value engineering, as discussed earlier, is a systematic approach to optimize the value of a project by identifying cost-effective alternatives. By analyzing the functions and objectives of the project, site managers can work collaboratively with the project team to propose innovative solutions that deliver the desired outcomes at a lower cost.

Material Selection: The choice of construction materials significantly impacts project costs. Site managers can seek cost-efficient alternatives by evaluating different materials based on their performance, durability, and cost-effectiveness. For instance, using recycled or locally sourced materials can often reduce costs without compromising quality.

Technology Integration: Embracing technology in construction processes can lead to significant cost savings. Adopting BIM for design and planning, using construction management software for scheduling, and employing drones for site inspections are examples of how technology can streamline operations and reduce expenses.

Equipment Utilization: Optimizing the use of construction equipment is essential to control costs. Site managers should ensure that equipment is adequately utilized and properly maintained to extend its lifespan and minimize downtime.

Labor Efficiency: Labor costs account for a significant portion of construction expenses. Efficient workforce allocation, proper training, and workflow optimization can enhance productivity and reduce labor-related costs.

Lean Construction Practices: Implementing lean construction principles can help identify and eliminate wasteful activities, leading to more efficient processes and cost reductions. Lean practices focus on maximizing value while minimizing resources, time, and effort.

Competitive Bidding: When subcontractors and suppliers are involved, site managers can seek competitive bids to secure favorable pricing for various services and materials. Properly evaluating bids and negotiating contracts can lead to cost-efficient partnerships.

Value-Driven Procurement: Instead of focusing solely on the lowest upfront cost, site managers should consider the long-term value and total cost of ownership when procuring materials and services. This approach ensures that investments yield sustainable benefits over the project's lifecycle.

Continuous Improvement: Encourage a culture of continuous improvement among the project team. Regularly assess project performance, review lessons learned from past projects, and implement best practices to enhance cost efficiency on future endeavors.

By diligently seeking cost-efficient alternatives and solutions, site managers can contribute significantly to the financial success of construction projects. It requires proactive thinking, collaboration with stakeholders, and a commitment to delivering projects that meet client expectations while adhering to budget constraints. Aspiring and current site managers who excel in this aspect of project management are more likely to build a reputation for delivering high-quality projects on time and within budget.

Balancing Quality and Cost Considerations

As a site manager, your primary challenge lies in striking the right balance between delivering a high-quality construction project and managing costs effectively. This equilibrium is crucial for ensuring project success and meeting the expectations of clients and stakeholders.

To achieve this balance, start by thoroughly understanding the project requirements. Clearly define the scope, objectives, and performance expectations in collaboration with the project team and stakeholders. This clarity will ensure that all decisions align with the desired outcomes.

Value engineering plays a significant role in finding the balance between quality and cost. Through careful analysis, site managers can identify areas where cost savings can be achieved without compromising the project's overall quality and functionality.

Performing a cost-quality trade-off analysis at various stages of the project is essential. This involves weighing potential cost savings against the impact on quality and performance. Data-driven decision-making will help determine the best approach to achieve the desired results within budget constraints.

Collaboration with the project team, including architects, engineers, subcontractors, and suppliers, is crucial in achieving the desired balance. Engage all stakeholders in discussions about cost implications and quality requirements, encouraging open communication to explore innovative solutions.

Consider the long-term implications of decisions related to quality and cost. Certain choices may have higher upfront costs but lead to lower maintenance and operational expenses over the project's lifecycle. Performing a lifecycle cost analysis will help identify cost-effective options with long-term benefits.

Identify critical elements of the project that significantly impact quality and safety. Prioritize spending on these elements while finding opportunities to economize in other areas. For example, investing in high-quality structural materials may be essential for safety, while other aesthetic elements could be cost-optimized.

Implement robust quality control measures to ensure that construction activities adhere to established standards. Effective quality control helps prevent costly rework and ensures that the final deliverables meet the expected quality levels.

Foster a culture of continuous improvement within the project team. Regularly review and evaluate performance data, including quality metrics and cost tracking. Use this data to identify areas for improvement and take proactive measures to address issues.

Recognize that certain cost-saving measures might introduce risks to the project's quality and timeline. Assess these risks and implement risk mitigation strategies to safeguard project success.

Balancing quality and cost considerations requires a comprehensive and strategic approach. It involves making informed decisions based on a deep understanding of project requirements, collaborating with stakeholders, and maintaining a focus on delivering value to clients. Aspiring and current site managers who excel in finding this balance are more likely to deliver successful construction projects that meet the highest quality standards within budget constraints.

CHAPTER 10: PROBLEM-SOLVING AND CONFLICT RESOLUTION

This chapter equips site managers with essential skills for effective problem-solving and conflict resolution. It emphasizes inclusive decision-making, active listening, and communication to tackle challenges and maintain a harmonious work environment. Preparedness, a positive attitude, and a problem-solving mindset are crucial for successful outcomes.

Identifying and Addressing On-Site Challenges

Site managers face numerous challenges during construction projects, and their ability to identify and address these issues promptly is crucial for project success. One of the first steps in tackling on-site challenges is to conduct a thorough project risk assessment before commencing work. This assessment helps identify potential hurdles that may arise during the construction process, such as weather conditions, labor shortages, material delays, or unforeseen ground conditions.

Once challenges are identified, site managers need to develop effective strategies to address them. This often involves collaborating with various stakeholders, including subcontractors, suppliers, and project owners. Communication plays a vital role in this process, as clear and transparent communication ensures that everyone involved is on the same page and understands their roles in resolving the challenges.

Problem-solving is an integral part of addressing on-site challenges. Site managers need to approach problems systematically and analytically. This involves gathering relevant data, analyzing the situation, and exploring various possible solutions. In some cases, it might be beneficial to involve the project team in the decision-making process to leverage diverse perspectives and expertise.

Conflict resolution is another critical aspect of addressing on-site challenges. Conflicts can arise due to differences in opinions, interests, or priorities among team members or stakeholders. Site managers should employ conflict resolution techniques such as active listening, mediation, and negotiation to resolve conflicts amicably.

Maintaining a positive work culture is also essential when dealing with challenges. A motivated and supportive team is better equipped to tackle problems collectively and find creative solutions. Site managers should foster a culture of open communication, trust, and collaboration to keep the team engaged and focused on overcoming obstacles.

Furthermore, learning from past experiences can help site managers better prepare for future challenges. Keeping detailed records of previous projects, including both successes and failures, enables the team to learn from their mistakes and implement improvements for future endeavors.

Overall, the ability to identify, address, and overcome on-site challenges is a fundamental skill for site managers. With effective problem-solving, conflict resolution, and a positive work culture, site managers can navigate through difficulties and keep their construction projects on track to successful completion.

Analyzing Common Construction Challenges and Pitfalls

Construction projects often encounter various challenges and pitfalls that can impact the project's timeline, budget, and overall success. Aspiring and current site managers must be well-prepared to tackle these issues effectively. Below are some common construction challenges and pitfalls that site managers may face:

Weather Conditions: Adverse weather, such as heavy rain, snow, or extreme heat, can cause delays in construction activities and affect worker safety. Site managers should monitor weather forecasts regularly and plan accordingly to minimize weather-related disruptions.

Labor Shortages: Finding skilled and reliable labor can be challenging, especially during peak construction periods. Site managers should focus on workforce planning and recruitment strategies to ensure they have a sufficient workforce to meet project demands.

Material Delays and Shortages: Delayed or inadequate delivery of construction materials can significantly impact the project's progress. Site managers must maintain effective communication with suppliers and have backup plans in place to handle material shortages.

Unforeseen Ground Conditions: Site managers may encounter unexpected ground conditions, such as poor soil quality or hidden underground utilities, which can lead to project delays and additional costs. Conducting thorough site investigations and soil tests before starting construction can help mitigate these risks.

Permitting and Regulatory Compliance: Obtaining necessary permits and complying with regulations can be time-consuming and complex. Site managers must stay updated with local building codes and ensure all necessary approvals are in place before commencing work.

Safety Incidents and Accidents: Safety should be a top priority on construction sites. Site managers must implement strict safety protocols, conduct regular safety training, and enforce safety practices to prevent accidents and injuries.

Scope Creep: Changes to project scope can occur due to client requests or design modifications. Site managers should carefully assess the impact of scope changes on the project's timeline and budget and communicate any adjustments to stakeholders.

Communication Breakdown: Poor communication between team members, subcontractors, and stakeholders can lead to misunderstandings and delays. Site managers should foster open and clear communication channels to ensure everyone is aligned with project objectives.

Budget Overruns: Managing project costs is critical to the project's financial success. Site managers should closely monitor expenses, track budget variances, and implement cost-control measures to avoid exceeding the allocated budget.

Inadequate Planning and Scheduling: Insufficient project planning and scheduling can lead to inefficiencies and delays. Site managers should create detailed project plans, establish realistic timelines, and regularly review progress to ensure the project stays on track.

Quality Control Issues: Failing to maintain high-quality standards can result in rework and additional costs. Site managers must implement robust quality control processes and conduct inspections to ensure work meets the specified standards.

Contractual Disputes: Disagreements with subcontractors or suppliers over contractual terms can lead to conflicts and delays. Site managers should ensure that all contracts are clear and detailed to minimize the risk of disputes.

By analyzing and understanding these common challenges and pitfalls, site managers can proactively develop strategies to mitigate risks and improve project outcomes. Being proactive, communicative, and adaptable are key traits that can help site managers navigate through challenges and ensure successful construction projects.

Developing Problem-Solving Techniques for Site Managers

As a site manager, developing strong problem-solving techniques is essential for effectively addressing the myriad challenges that can arise during construction projects. Problem-solving skills enable site managers to make informed decisions, overcome obstacles, and ensure successful project outcomes.

Analytical thinking is a valuable skill that involves breaking down complex problems into smaller, manageable parts and analyzing each component to gain a deeper understanding. Site managers should carefully examine the root causes of issues to identify the most effective solutions.

Data-driven approaches are becoming increasingly important in construction management. Collecting and analyzing data can provide valuable insights into project performance and potential problems. Site managers should utilize project management software and data analytics tools to make data-driven decisions.

Collaborative problem-solving is a powerful approach that encourages team members, subcontractors, and stakeholders to work together to find solutions. Regular meetings and open discussions can lead to innovative ideas and a more engaged project team.

Creative thinking is another essential skill for site managers. Embracing creative thinking allows them to explore unconventional solutions to problems. Thinking outside the box can lead to breakthrough ideas that optimize project efficiency and outcomes.

Risk management is crucial for proactive problem-solving. Site managers should conduct risk assessments and put contingency plans in place to mitigate the impact of unforeseen events.

Establishing decision-making frameworks helps site managers approach problem-solving in a systematic manner. Tools like cost-benefit analysis and SWOT (Strengths, Weaknesses, Opportunities, Threats) analysis can aid in making well-informed decisions.

Continuous learning is vital for staying updated on industry trends, best practices, and new technologies. It equips site managers with valuable knowledge and enhances their problem-solving capabilities.

Conflict resolution skills are essential for resolving disputes among team members and stakeholders. Site managers should develop effective communication and mediation skills to maintain positive working relationships.

Flexibility and adaptability are critical traits for site managers. Construction projects are dynamic, and unexpected challenges are inevitable. Site managers should be prepared to adjust plans and strategies as needed to navigate changing circumstances.

Seeking advice from subject matter experts can provide valuable guidance and insights when facing complex challenges. Engaging with experienced consultants or industry peers can offer fresh perspectives.

Prioritization is key for efficient problem-solving. Site managers must prioritize issues based on their impact on the project's critical path and overall objectives.

Reviewing past projects and analyzing lessons learned can help site managers avoid repeating mistakes and implement effective solutions in future endeavors.

By honing these problem-solving abilities, aspiring and current site managers can confidently tackle challenges, foster continuous improvement, and deliver successful construction projects.

Creating Contingency Plans for Unforeseen Issues
Creating comprehensive contingency plans is a crucial aspect of effective construction project management. Contingency plans are developed to anticipate and address unforeseen issues that may arise during the course of a project. By proactively planning for potential challenges, site managers can minimize disruptions, maintain project schedules, and ensure successful project delivery.

The first step in creating a contingency plan is to conduct a thorough risk assessment. Site managers should identify and analyze potential risks that could impact the project, such as adverse weather conditions, material shortages, labor disruptions, or design changes. Understanding the specific risks allows them to develop targeted and effective contingency strategies.

Once the risks are identified, site managers should prioritize them based on their potential impact on the project's timeline, budget, and overall success. By focusing on high-impact risks, managers can allocate resources more efficiently and develop contingency plans for the most critical challenges.

Next, site managers should work collaboratively with their project team, stakeholders, and subcontractors to brainstorm and develop contingency solutions. This collaborative approach ensures that the plan is comprehensive and considers various perspectives and expertise.

Contingency plans should outline clear and detailed steps to be taken if specific risks materialize. These steps may include alternative sourcing strategies for materials, backup labor arrangements, adjusting project schedules, or implementing alternative construction methods. Each contingency measure should be well-defined, feasible, and aligned with the project's objectives.

Timelines and triggers for activating contingency plans should be established in advance. Site managers should continuously monitor the project's progress and risk factors to recognize when it becomes necessary to implement the contingency plan. Early recognition and swift action can prevent potential issues from escalating.

Contingency plans should be regularly reviewed and updated throughout the project's lifecycle. As circumstances change and new risks emerge, adjustments to the contingency plan may be required to ensure its relevance and effectiveness.

Additionally, communication is paramount in implementing contingency plans. Site managers must ensure that all relevant stakeholders are aware of the plan, their respective roles, and the steps to be taken if a contingency is activated. Effective communication ensures a cohesive response to unexpected challenges.

Contingency plans should not be viewed as a last resort but as a proactive strategy to enhance project resilience. By integrating contingency planning into the overall project management process, site managers can build a strong foundation for successful project delivery and demonstrate their commitment to risk management and project excellence.

Creating contingency plans for unforeseen issues is an integral part of construction project management. These plans enable site managers to respond effectively to challenges, minimize disruptions, and safeguard the project's success. By conducting thorough risk assessments, prioritizing risks, fostering collaboration, and maintaining communication, site managers can develop robust contingency plans that add value and resilience to their projects.

Resolving Conflicts among Team Members

Resolving conflicts among team members is an essential skill for site managers. Construction projects can be complex and demanding, involving various stakeholders with diverse perspectives and interests. Conflicts may arise due to differences in communication styles, work methods, or even personal issues. As a site manager, it is crucial to address these conflicts promptly and effectively to maintain a productive and harmonious work environment.

One of the first steps in conflict resolution is to actively listen to all parties involved. Site managers should create a safe and non-judgmental space for team members to express their concerns and perspectives openly. By listening attentively, managers can gain a deeper understanding of the root cause of the conflict and identify potential solutions.

Once the issues have been identified, the site manager should facilitate open and constructive communication between the conflicting parties. Encouraging team members to communicate directly with each other can help to clarify misunderstandings and promote empathy. As a mediator, the site manager should remain neutral and focus on finding common ground and mutually beneficial resolutions.

In some cases, conflicts may require a more structured approach to resolution. Site managers can facilitate conflict resolution workshops or team-building exercises to improve communication and collaboration among team members. These activities can foster a sense of camaraderie and trust, which can significantly reduce the likelihood of future conflicts.

In situations where conflicts are more challenging to resolve, it may be necessary for the site manager to intervene and provide guidance. This could involve setting clear expectations for behavior and performance, mediating discussions, or implementing conflict resolution techniques.

Conflict resolution should be approached with fairness and transparency. Site managers should avoid taking sides and ensure that decisions are based on objective criteria and the best interests of the project and the team as a whole.

Another vital aspect of conflict resolution is to follow up on the implemented solutions. Site managers should monitor the progress and effectiveness of the resolutions and be ready to make adjustments if needed. This ongoing support can help to prevent conflicts from resurfacing and build a culture of open communication and collaboration.

Additionally, site managers should be proactive in promoting a positive work culture that values respect, teamwork, and effective communication. Creating an environment where conflicts are addressed promptly and constructively can significantly reduce their occurrence and impact on the project.

Recognizing Sources of Conflicts in Construction Projects

Recognizing sources of conflicts in construction projects is essential for site managers to effectively address and mitigate potential issues. Construction projects involve various stakeholders, including owners, contractors, subcontractors, engineers, architects, and laborers, each with their own objectives and priorities. Understanding the potential sources of conflicts can help site managers be proactive in managing these situations before they escalate.

One common source of conflict in construction projects is unclear project goals and expectations. When project objectives, timelines, or deliverables are not well-defined, misunderstandings can arise, leading to disputes among team members. Site managers should ensure that project requirements are communicated clearly and that all parties involved have a shared understanding of the project's scope and objectives.

Differences in communication styles and ineffective communication can also contribute to conflicts. Misinterpretation of information, lack of proper communication channels, and failure to convey critical details can lead to misunderstandings and conflicts. Site managers must establish effective communication processes and encourage open and transparent communication among team members.

Moreover, resource constraints and scheduling issues can create tensions in construction projects. Limited availability of labor, equipment, or materials may lead to delays or disruptions in the project, causing frustration among stakeholders. Site managers should carefully plan and allocate resources to prevent conflicts arising from resource constraints.

Changes in project scope or design alterations can also be sources of conflicts. Change orders may affect project timelines, costs, and resource allocation, leading to disagreements between the owner, contractors, and subcontractors. Site managers should proactively manage change orders, communicate their implications clearly, and seek to find collaborative solutions.

Additionally, conflicting interests and priorities among project stakeholders can create tensions. For example, contractors may prioritize completing the project quickly, while the owner may prioritize cost containment. Balancing these conflicting interests requires effective negotiation and compromise from site managers.

Furthermore, differing work cultures and practices among team members can lead to conflicts. In diverse project teams, cultural differences, work habits, and communication styles may clash, resulting in misunderstandings and frictions. Site managers should promote cultural awareness and foster a collaborative work environment that values diversity and inclusion.

Lastly, unforeseen external factors, such as weather conditions, regulatory changes, or economic fluctuations, can introduce uncertainties and potential conflicts into the project. Site managers must be adaptable and responsive to these external influences to minimize their impact on the project.

Implementing Conflict Resolution Strategies
Implementing effective conflict resolution strategies is a critical skill for site managers to maintain a harmonious and productive construction environment. Conflicts are inevitable in any construction project due to the complexity of the tasks involved and the diversity of stakeholders. However, how conflicts are managed can significantly impact the project's success and team morale.

One essential conflict resolution strategy is active listening. When conflicts arise, it is crucial for site managers to listen attentively to the concerns of all parties involved. Actively listening allows site managers to gain a deeper understanding of the root causes of the conflict and the emotions behind the issues. By demonstrating empathy and understanding, site managers can create a more conducive atmosphere for resolving conflicts.

Another effective strategy is maintaining open communication. Site managers should encourage open and transparent communication among team members, promoting an environment where concerns and disagreements can be discussed openly and respectfully. Effective communication can prevent conflicts from escalating and enable early intervention in case of emerging issues.

Mediation is a valuable conflict resolution technique that site managers can utilize. In cases where conflicts persist despite initial attempts to address them, a neutral third party, such as an experienced mediator, can help facilitate the resolution process. Mediators can guide discussions, ensure that all voices are heard, and help parties find common ground for resolving their differences.

Collaborative problem-solving is an empowering conflict resolution approach. Site managers can bring conflicting parties together to collaboratively identify solutions to the issues at hand. By involving all stakeholders in the decision-making process, site managers can foster a sense of ownership and commitment to the resolution.

Negotiation is a fundamental skill for site managers in resolving conflicts. Negotiating involves finding mutually acceptable solutions that meet the needs and interests of all parties involved. Skilled negotiation can help reach compromises and resolve conflicts in a win-win manner.

Sometimes, conflicts may require more formal approaches, such as arbitration or adjudication. These methods involve the intervention of an external authority to make a binding decision on the dispute. While more formal, these processes can provide a final resolution to complex conflicts.

Conflict prevention is as crucial as resolution. Site managers should proactively identify potential sources of conflicts and implement measures to prevent them from arising. Clear communication of project objectives, roles, and responsibilities can help set expectations and minimize misunderstandings.

Lastly, documenting conflicts and their resolutions is essential for learning and improvement. Keeping a record of conflicts, resolutions, and lessons learned can guide future projects and help site managers develop strategies to prevent similar issues in the future.

Mediating Disputes and Promoting Team Harmony

Mediating disputes and promoting team harmony are vital skills that site managers must possess to maintain a productive and cohesive work environment on construction projects. Conflicts and disagreements among team members are inevitable, given the complexity of construction tasks and the diverse group of individuals involved. As a site manager, it is crucial to handle disputes professionally and effectively to avoid disruptions and foster a positive team dynamic.

Mediation is an essential conflict resolution technique that site managers can use to address disputes between team members. As a neutral third party, the site manager can facilitate discussions between the involved parties, allowing them to express their perspectives and concerns in a safe and controlled environment. The goal of mediation is to help team members find common ground, reach mutual understanding, and identify collaborative solutions to their differences. By actively listening to all parties and refraining from taking sides, site managers can build trust and credibility as mediators.

Effective communication is the foundation of successful mediation. Site managers should encourage open and honest dialogue among team members, fostering an atmosphere of respect and understanding. By promoting active listening and ensuring that all voices are heard, site managers can create an environment where conflicts can be addressed constructively.

During mediation, it is essential for site managers to remain impartial and unbiased. The focus should be on the interests of the parties involved and finding solutions that benefit the entire team. By adopting a problem-solving approach and encouraging compromise, site managers can help team members move past their differences and focus on the project's objectives.

Site managers can proactively work to prevent conflicts from escalating by implementing conflict resolution procedures and promoting a culture of cooperation and respect within the team. Regular team-building activities and open forums for discussing concerns can help foster strong working relationships and prevent conflicts from arising.

In addition to mediating disputes between team members, site managers may also need to address conflicts involving external stakeholders, such as clients, subcontractors, or regulatory authorities. In these situations, effective communication and negotiation skills become even more crucial. The ability to diplomatically handle disagreements and find mutually beneficial solutions is essential in maintaining positive relationships with stakeholders and ensuring the smooth progress of the project.

Promoting team harmony is not only about resolving conflicts but also about creating a supportive and inclusive work environment. Site managers should recognize and acknowledge the contributions of team members, foster a culture of collaboration, and provide opportunities for professional growth and development.

Managing Unforeseen Circumstances and Changes

As site managers, one of the most significant challenges we face is managing unforeseen circumstances and changes that can arise during construction projects. Despite careful planning and preparation, the dynamic nature of the construction industry means that unexpected events and situations can occur, requiring quick and effective responses. Aspiring and current site managers must develop the skills and strategies to handle these challenges with professionalism and adaptability.

One of the first steps in managing unforeseen circumstances is to establish a robust risk management process. This involves identifying potential risks and uncertainties that may impact the project and developing contingency plans to mitigate their effects. By anticipating possible challenges and having a plan in place, site managers can respond more effectively when unexpected events occur.

In the face of unforeseen changes, site managers must be prepared to make decisions swiftly and decisively. It is crucial to gather all available information, analyze the potential impacts, and consult with relevant stakeholders before making informed choices. Effective decision-making ensures that projects can continue to progress without unnecessary delays or disruptions.

Maintaining open lines of communication with all stakeholders is essential when managing unforeseen circumstances and changes. Regular and transparent communication allows site managers to keep all parties informed of developments and potential impacts. It also fosters a collaborative approach to problem-solving, as stakeholders can contribute their insights and suggestions to find the best solutions.

Flexibility is a key attribute for site managers dealing with unforeseen circumstances. Projects rarely go exactly as planned, and the ability to adapt to changing situations is crucial for success. By being open to alternative approaches and willing to adjust plans as needed, site managers can navigate through challenges more effectively.

When unforeseen circumstances require changes to the project scope, timeline, or budget, it is essential to document these changes thoroughly. Proper documentation provides a record of the decisions made and the reasoning behind them, which can be invaluable for future reference and potential dispute resolution.

In some cases, unforeseen circumstances may require additional resources or expertise. Site managers should be proactive in seeking support from subject matter experts or external consultants if necessary. Collaboration with experts can bring valuable insights and solutions to complex challenges.

It is also vital for site managers to remain calm and composed under pressure. Construction projects can be stressful, especially when faced with unexpected events. By maintaining a positive attitude and demonstrating strong leadership, site managers can instill confidence in the team and maintain focus on achieving project goals.

Lastly, learning from past experiences is crucial for improving future project management. After successfully managing unforeseen circumstances, site managers should conduct a post-mortem analysis to identify lessons learned and areas for improvement. This continuous learning approach ensures that each project becomes an opportunity for growth and refinement of skills.

Adapting to Unexpected Situations and Project Changes

Adapting to unexpected situations and project changes is a critical skill for successful project delivery in the construction industry. Construction projects are inherently complex and subject to a wide range of external factors that can introduce unforeseen challenges. Aspiring and current site managers must develop a mindset of flexibility and resilience to navigate through these changes effectively.

One of the key aspects of adapting to unexpected situations is maintaining a proactive approach to risk management. This involves identifying potential risks early in the project, analyzing their potential impact, and developing contingency plans to address them. By anticipating possible challenges, project teams can be better prepared to respond swiftly and mitigate the effects of unexpected events.

Effective communication is another critical factor in adapting to unexpected situations. Keeping all stakeholders informed of developments, challenges, and potential changes is essential for maintaining a collaborative and cohesive project environment. Regular and transparent communication fosters trust among team members and stakeholders, which is crucial for resolving issues and making collective decisions.

When unexpected situations arise, project teams must be ready to make quick and well-informed decisions. This requires gathering relevant information, consulting with experts or team members, and analyzing the potential consequences of each course of action. Decisiveness is vital to avoid delays and keep the project on track.

Flexibility in project planning and execution is crucial when facing unexpected situations and project changes. Rigidity can hinder progress and lead to inefficiencies. Project teams must be open to adapting project schedules, resource allocations, and methodologies to accommodate new information and developments.

Moreover, being adaptable also means embracing innovation and new technologies. Modern construction practices often offer solutions to unexpected challenges. Staying up-to-date with industry advancements can provide project teams with innovative approaches to problem-solving and improve overall project performance.

Collaboration and teamwork play a significant role in adapting to unexpected situations. Encouraging open discussions and input from all team members can lead to creative solutions and foster a sense of ownership and responsibility among the team.

Project teams must also be mindful of the impact of unexpected changes on project costs and budgets. Regular monitoring of project finances can help identify potential cost overruns and allow for adjustments to keep expenses within acceptable limits.

It is essential for project teams to maintain a positive and proactive attitude in the face of unexpected situations. Challenges are a part of every construction project, and how they are handled can greatly impact project outcomes. By approaching challenges with a problem-solving mindset and an optimistic outlook, project teams can inspire confidence and trust among stakeholders.

Finally, learning from past experiences is vital for improving adaptability. After each project, project teams should conduct a post-mortem analysis to identify lessons learned and areas for improvement. This continuous learning process ensures that project teams can continually refine their skills and approaches to better handle unexpected situations in future projects.

Implementing Change Management Procedures

Implementing change management procedures is a crucial aspect of effective construction project management. Change is inevitable in construction projects due to various factors such as evolving client requirements, design modifications, unforeseen challenges, and external influences. Aspiring and current site managers must have a well-defined change management process in place to ensure that changes are properly assessed, approved, and implemented without compromising project objectives and timelines.

The first step in implementing change management procedures is to establish a formal change request process. This process should outline the steps for submitting change requests, including the required documentation and justification for the proposed change. It should also define the roles and responsibilities of team members involved in reviewing and approving change requests.

Once a change request is received, it should undergo a thorough assessment to evaluate its impact on the project. This assessment should consider factors such as cost implications, schedule adjustments, potential risks, and effects on project scope and quality. The change request should be evaluated against the project's overall goals and objectives to determine its alignment with the project's vision.

After the assessment, the change request should be reviewed by a designated change control board or committee. This board should consist of key stakeholders, including project owners, site managers, architects, and engineers. Their role is to assess the change request objectively and make informed decisions based on its merits and alignment with project goals.

Effective communication is vital throughout the change management process. All stakeholders involved in the change request should be kept informed of its status and any decisions made. Clear and transparent communication helps build trust and ensures that everyone is on the same page regarding the changes being implemented.

It is essential to document all change management decisions and actions. This documentation serves as a historical record of the changes made during the project and provides a reference for future evaluations. Proper documentation also helps in identifying trends and patterns related to changes, which can inform decision-making in similar future projects.

Risk management is an integral part of change management. As part of the change assessment process, project teams should identify potential risks associated with the proposed change and develop mitigation strategies to address them. This proactive approach ensures that risks are considered and managed effectively during the change implementation.

It is essential to monitor and track the impact of approved changes on the project. Regular project status updates should include information on change implementation progress, associated costs, and any deviations from the original project plan. Monitoring helps to identify any emerging issues and allows for timely corrective actions if needed.

Change management procedures should align with the overall project management framework and integrate seamlessly into existing processes. The goal is to streamline change management to make it a natural part of the project lifecycle rather than a burdensome and disruptive process.

Moreover, fostering a positive culture towards change is crucial for successful implementation. Team members should be encouraged to embrace change as an opportunity for improvement and growth. When stakeholders understand the rationale behind changes and see their benefits, resistance to change is reduced, and the overall change management process becomes more effective.

Mitigating Risks and Minimizing Project Disruptions

Mitigating risks and minimizing project disruptions are critical responsibilities for site managers in construction projects. Risks are inherent in construction projects due to the complex and dynamic nature of the industry. Unforeseen events, changes in scope, adverse weather conditions, material shortages, and unforeseen regulatory requirements can disrupt project schedules and budgets. Aspiring and current site managers must proactively identify potential risks, develop mitigation strategies, and implement measures to minimize disruptions throughout the project lifecycle.

One of the first steps in risk mitigation is conducting a comprehensive risk assessment. This involves identifying potential risks that could impact the project, analyzing their likelihood and potential consequences, and prioritizing them based on their severity. The risk assessment should involve key stakeholders and subject matter experts to ensure a comprehensive and informed analysis.

Once the risks are identified, site managers should develop

mitigation strategies to address each risk. These strategies may include contingency plans, alternative approaches, and measures to avoid or transfer risks. Contingency plans are crucial in dealing with unexpected events, providing a roadmap for action if risks materialize.

Effective risk mitigation requires open and transparent communication among all project stakeholders. Site managers should proactively communicate with project owners, subcontractors, suppliers, and other team members to ensure everyone is aware of potential risks and the corresponding mitigation measures. Regular project status updates and risk reviews help keep stakeholders informed and engaged in the risk management process.

In addition to risk mitigation, site managers must be proactive in addressing project disruptions as they arise. When disruptions occur, prompt action is necessary to minimize their impact on project schedules and budgets. This may involve reallocating resources, adjusting timelines, or renegotiating contracts with suppliers and subcontractors.

Site managers should also be vigilant in monitoring project progress and identifying potential disruptions before they escalate. Early detection allows for timely intervention and mitigation, preventing minor issues from snowballing into major problems.

An essential aspect of minimizing project disruptions is fostering a culture of safety and quality on the construction site. By prioritizing safety measures, adhering to industry standards, and implementing strict quality control processes, site managers can reduce the likelihood of accidents, rework, and delays.

Furthermore, having a skilled and well-trained workforce is crucial for risk mitigation and minimizing disruptions. Site managers should invest in training and development programs to enhance the skills and capabilities of their teams, thereby increasing the overall efficiency and productivity of the project.

Collaboration and coordination among all project stakeholders are vital in risk mitigation and disruption minimization. Site managers should foster strong relationships with subcontractors, suppliers, and other team members to ensure seamless coordination and a shared commitment to project success.

CONCLUSION

Key Takeaways for Becoming an Effective Site Manager

Becoming an effective site manager requires a combination of technical skills, leadership abilities, and strong communication. Here are some key takeaways for aspiring and current site managers to excel in their roles:

Technical Expertise: Site managers should have a solid understanding of construction processes, project management principles, and industry regulations. Continuous learning and staying updated with the latest advancements in the field are essential.

Leadership and Team Management: Site managers must be effective leaders who can motivate and inspire their teams. Building a positive and collaborative work environment fosters productivity and ensures the successful completion of projects.

Effective Communication: Clear and open communication is crucial for site managers to convey expectations, address concerns, and ensure everyone is on the same page. Regular meetings and progress updates with stakeholders are vital.

Problem-Solving Skills: Construction projects come with challenges. Site managers must be adept at identifying issues, analyzing options, and implementing effective solutions to keep projects on track.

Time and Resource Management: Efficiently managing time and resources is vital for meeting project milestones and budgets. Site managers should prioritize tasks, allocate resources wisely, and adapt to changing project needs.

Safety First: Safety should be the top priority on construction sites. Site managers must enforce strict safety protocols and promote a culture of safety among workers and subcontractors.

Quality Control: Ensuring high-quality workmanship and materials is essential for the success and longevity of the project. Site managers should implement rigorous quality control processes.

Adapting to Change: Construction projects are dynamic, and unexpected changes are inevitable. Site managers must be flexible and adapt to new situations while maintaining project objectives.

Conflict Resolution: Conflicts can arise among team members or stakeholders. Effective site managers should be skilled in resolving disputes and maintaining positive relationships.

Risk Management: Identifying potential risks, developing mitigation plans, and monitoring risks throughout the project are crucial to minimize disruptions and avoid costly setbacks.

Financial Acumen: Site managers should be proficient in budgeting, cost control, and financial forecasting to ensure the project stays within budget and achieves financial objectives.

Continuous Improvement: Embrace a mindset of continuous improvement, seeking feedback, and learning from past projects to refine management strategies and enhance overall performance.

By applying these key takeaways, aspiring and current site managers can navigate the challenges of construction projects confidently and achieve successful outcomes while ensuring the safety, quality, and efficiency of the entire construction process.

Recapitulating Essential Skills and Knowledge for Site Managers
Emphasizing the Importance of Leadership and Adaptability
Encouraging Continuous Learning and Professional Growth

Continuous Learning and Professional Development in Site Management

Continuous learning and professional development are crucial aspects of becoming a successful site manager. The construction industry is constantly evolving, with new technologies, regulations, and best practices emerging regularly. To stay at the forefront of the industry and enhance their skills, aspiring and current site managers must embrace a commitment to lifelong learning. Here are some key points to consider:

Industry Knowledge: Site managers should dedicate time to stay updated with industry trends, new construction methods, and technological advancements. Engaging in industry conferences, workshops, and webinars can provide valuable insights and networking opportunities.

Certifications and Training: Pursuing relevant certifications, such as Project Management Professional (PMP) or Construction Safety certifications, demonstrates a commitment to professional growth and enhances credibility among peers and stakeholders.

Mentorship: Seeking guidance from experienced site managers or construction professionals can provide valuable insights and practical knowledge. Mentorship can help aspiring site managers navigate challenges and make informed decisions.

Networking: Building a strong professional network within the construction industry opens doors to new opportunities, collaboration, and knowledge-sharing. Networking events, industry associations, and online forums are excellent platforms for expanding connections.

Continuing Education: Enrolling in courses or workshops related to construction management, leadership, safety, and other relevant topics helps site managers acquire specialized skills and knowledge.

Reading and Research: Regularly reading industry publications, books, and research articles can provide valuable insights and keep site managers informed about the latest developments and best practices.

Soft Skills Development: Effective site managers should focus on improving their communication, negotiation, and conflict resolution skills. These soft skills are vital for building strong relationships with team members, stakeholders, and clients.

Technology Adoption: Embracing construction management software, Building Information Modeling (BIM), and other technological tools enhances efficiency and productivity in project management.

Seek Feedback: Encouraging feedback from team members and stakeholders helps site managers identify areas for improvement and gain insights into their leadership style.

Reflect and Apply Learnings: Regularly reflect on project experiences and apply the lessons learned from successes and challenges to future projects. This reflection process is vital for continuous improvement.

Attend Seminars and Workshops: Participating in seminars and workshops on various construction-related topics helps site managers gain fresh perspectives and new ideas for problem-solving.

Empower Team Members: Encourage team members to pursue professional development and provide opportunities for their growth. A knowledgeable and skilled team contributes to the overall success of the projects.

Continuous learning and professional development are essential for site managers to thrive in the construction industry. By staying informed about the latest trends, enhancing their technical and soft skills, and fostering a culture of learning within their teams, site managers can lead projects with confidence, efficiency, and effectiveness.

Identifying Learning Opportunities and Resources

As a site manager, identifying learning opportunities and utilizing available resources is critical for continuous growth and professional development. Here are some strategies to identify and leverage learning opportunities:

Industry Associations and Conferences: Joining construction-related associations and attending conferences provides access to workshops, seminars, and networking events that offer valuable insights into industry trends and best practices.

Online Courses and Webinars: Numerous platforms offer online courses and webinars on various construction topics, ranging from project management to safety and sustainability. These flexible learning options allow site managers to enhance their skills at their convenience.

Books and Publications: Reading books, industry publications, and construction journals keeps site managers updated with the latest knowledge and research in the field. Many reputable authors and organizations publish informative content on construction management.

Workplace Training Programs: Encourage team members to

attend training programs conducted by the company or external providers. In-house training on specific topics can be tailored to address the organization's unique needs.

Certification Programs: Pursuing certifications relevant to construction management, safety, or sustainability provides formal recognition of skills and knowledge and enhances professional credibility.

Peer Learning: Engaging in discussions and knowledge-sharing with other site managers and construction professionals can offer valuable insights and practical solutions to common challenges.

Online Forums and Discussion Groups: Participating in online construction forums and discussion groups provides an opportunity to learn from the experiences and expertise of others in the industry.

Company Mentorship Programs: Establishing a mentorship program within the organization allows aspiring site managers to receive guidance from experienced leaders, fostering professional growth.

Visiting Construction Sites: Organizing site visits to ongoing projects, especially those executed by industry leaders, can offer valuable exposure to innovative construction techniques and project management practices.

University and College Courses: Enrolling in construction-related courses at universities or colleges can provide formal education and a deeper understanding of construction principles.

Government and Industry Reports: Studying government reports and industry analyses helps site managers stay informed about regulatory changes, market trends, and emerging opportunities.

Podcasts and Videos: Listening to construction-related podcasts or watching informative videos can be an engaging way to gain knowledge while on the move.

By proactively seeking learning opportunities and utilizing various resources available, site managers can continuously expand their expertise, enhance their decision-making abilities, and lead construction projects more effectively. Embracing a growth mindset and a commitment to lifelong learning will ultimately contribute to professional success and advancement within the construction industry.

Pursuing Professional Certifications and Designations

Pursuing professional certifications and designations is an essential step for site managers who seek to establish themselves as experts in their field and demonstrate their commitment to excellence. These certifications validate their skills, knowledge, and experience, enhancing their credibility and opening up new career opportunities. Here are some notable certifications and designations that aspiring and current site managers can consider:

Project Management Professional (PMP): This globally recognized certification focuses on project management principles and practices, applicable in various industries and countries.

Certified Construction Manager (CCM): A designation that showcases expertise in construction management, widely applicable to construction projects globally.

Construction Safety Certification: Various organizations offer safety certifications that demonstrate proficiency in ensuring safe work environments on construction sites, relevant to international safety standards.

LEED Accredited Professional (LEED AP): A certification that indicates expertise in sustainable and eco-friendly construction practices, applicable to green building projects worldwide.

Certified Cost Professional (CCP): A certification that validates competence in cost engineering and cost management, useful for construction projects in different countries.

Construction Documents Technologist (CDT): A designation that signifies understanding of construction documentation and contract administration principles, relevant globally.

Certified Professional Constructor (CPC): A certification that demonstrates construction management competence, suitable for construction projects in various regions.

BIM Certifications: Certifications that emphasize expertise in digital modeling and collaboration, applicable to construction projects using BIM worldwide.

Occupational Health and Safety Certifications: Various safety certifications, such as Certified Safety Professional (CSP) or Certified Health and Safety Specialist (CHSS), demonstrate commitment to safety standards applicable across different countries.

Contract Management Certifications: Certifications in contract management equip site managers with valuable skills for handling contractual arrangements in diverse international contexts.

These certifications and designations provide site managers with a globally recognized skill set, enabling them to excel in their roles regardless of the country or region where they work. As construction projects continue to be executed worldwide, having these certifications can elevate a site manager's capabilities and marketability in the industry.

Cultivating a Culture of Lifelong

Learning in Site Management

Cultivating a culture of lifelong learning is essential for site managers to stay updated with industry advancements, enhance their skills, and adapt to changing project demands. Aspiring and current site managers should prioritize continuous learning and foster an environment that encourages professional development among their teams.

Lead by example by actively engaging in professional development opportunities, attending workshops, conferences, and pursuing certifications. Demonstrating a commitment to learning inspires their team members to do the same.

Allocate resources and time for regular training sessions and workshops to address specific skill gaps or introduce new technologies and methodologies. This can be done both in-house and by partnering with external training providers.

Foster a supportive atmosphere where team members can discuss their career goals and aspirations. Encourage them to create personalized development plans, aligning their objectives with the organization's goals.

Establish regular knowledge-sharing sessions or lunch-and-learn events where team members can share best practices, lessons learned from projects, and insights from recent training.

Utilize e-learning platforms and online courses to provide accessible and flexible learning opportunities for team members. These platforms offer a wide range of topics, catering to various interests and skills.

Develop a library or resource center with books, journals, and industry publications related to construction, project management, safety, and relevant disciplines. Encourage team members to utilize these resources.

Acknowledge team members who pursue professional development by celebrating their achievements and contributions to the team's growth.

Incorporate learning and development objectives into performance evaluations to highlight the importance of continuous improvement and incentivize learning efforts.

Implement mentorship and coaching initiatives, pairing experienced site managers with newer team members to foster knowledge transfer and personal growth.

Encourage team members to stay updated on the latest construction trends, regulations, and technologies through networking events, webinars, and industry publications.

By fostering a culture of lifelong learning, site managers can enhance their team's capabilities, improve project outcomes, and create a dynamic and adaptive work environment. Embracing continuous learning ensures that site managers and their teams remain at the forefront of industry developments, fostering innovation and success in their projects.

Recommended resources for further reading

1. "Construction Management: Principles and Practice" by Chris March and Stephen Emmitt - This comprehensive book covers various aspects of construction management, including project planning, procurement, and risk management.

2. "Project Management for Construction: Fundamental Concepts for Owners, Engineers, Architects, and Builders" by Chris Hendrickson and Tung Au - This book provides a solid foundation in project management principles and practices specifically tailored to the construction industry.

3. "The Construction Project Management Success Guide"

by Andreas P. - A practical guide that offers tips and techniques for successful project management in construction.

4. "Construction Project Management: A Complete Introduction" by Alison Dykstra - An introductory guide to construction project management, covering key concepts and best practices.

5. "Site Management for Engineers" by J. Paul Guyer - A comprehensive guide specifically focused on the responsibilities and challenges of site managers in engineering projects.

6. "Project Management in Construction" by Sidney M. Levy - This book delves into project management techniques with a specific focus on construction projects.

7. "Construction Operations Manual of Policies and Procedures" by Andrew Civitello Jr. - A valuable resource for construction managers, offering practical guidelines for various construction processes and procedures.

8. "Construction Safety Management" by David L. Goetsch - A guide to safety management principles and practices in construction, emphasizing the importance of creating a safe work environment.

9. "Lean Construction Management: The Toyota Way" by Shang Gao, Sui Pheng Low, and Heng Li - This book explores how lean principles can be applied to construction management to improve efficiency and eliminate waste.

10. "The GPM Reference Guide to Green and

Sustainable   Building" by Reinhard Loske and Heather L. Dodd - A guide to integrating sustainable practices into construction projects, promoting environmentally responsible construction.

ABOUT THE AUTHOR

Steven Smith, Ph.d.

Steven Smith is a renowned expert in the field of Construction Management, with a wealth of knowledge and experience spanning both academia and industry. Holding a doctorate in Construction Management, Steven has dedicated his career to advancing the field and contributing to its body of knowledge.

Throughout his academic journey, Steven's passion for understanding the intricacies of construction processes and finding innovative solutions to industry challenges became evident. His doctoral research focused on optimizing project management practices and enhancing productivity in construction projects, leading to a profound understanding of various aspects of construction management and their impact on project success.

As an accomplished author, Steven Smith has made significant contributions to the construction industry through his various publications. Some of his notable works include:

"Construction Business Startup 101": A comprehensive guide that equips aspiring entrepreneurs with the essential tools and knowledge to successfully launch and manage a thriving construction business.

"Construction Management Blueprint": A blueprint that lays the foundation for effective construction project management, guiding professionals to orchestrate seamless projects from inception to completion.

"The Art of Construction Project and Business Management": A masterful exploration of the delicate balance between managing construction projects efficiently and nurturing a prosperous construction business.

"A Practical Approach to Risk Management in Construction": An indispensable resource that empowers construction professionals to identify, assess, and mitigate risks in construction projects, ensuring successful outcomes in the face of uncertainty.

"The Dictionary of Construction Terminologies": An invaluable reference, compiling a vast array of construction terminologies, ensuring clear communication and precise understanding of industry jargon.

Steven's commitment to sharing his wealth of expertise and supporting the success of construction projects worldwide is evident in his works. His books serve as essential resources for professionals and enthusiasts alike, showcasing Steven Smith as a leading authority in the construction industry. Through his contributions, he continues to enhance the practices of construction management and inspire future generations to excel in the realm of on-site operations.

BOOKS BY THIS AUTHOR

Construction Business Startup 101: Laying The Groundwork For Success

Embark on a transformative journey to construction entrepreneurship with this comprehensive guide. Discover actionable steps, real-world insights, and the power of innovation to thrive in the dynamic industry. Craft a compelling business plan, build a strong team, conquer financial challenges, and draw inspiration from successful professionals. Turn your vision into a thriving construction business and make a lasting impact in this competitive landscape. Unleash your potential and set forth on a path of growth, achievement, and fulfillment!

Construction Management Blueprint: A Comprehensive Guide To Successful Project Delivery

This book is the ultimate guide that equips you with the knowledge and tools to excel in the complex world of construction projects.

Covering key aspects of construction management, from project initiation to post-construction evaluation, it provides practical insights, real-life case studies, and invaluable tips for success.

Inside these pages, you'll find:

A step-by-step guide to project initiation and feasibility studies, helping you identify objectives, assess market demand, and engage stakeholders effectively.

In-depth coverage of project planning and design, including goal setting, scope definition, work breakdown structures, and sustainable design principles.

Extensive discussions on cost estimation techniques, budgeting, resource allocation, value engineering, and contract pricing and negotiation strategies.

Detailed insights into construction scheduling, resource procurement, site layout and logistics, and risk management, ensuring smooth project execution.

Thorough examinations of quality assurance and control, materials testing and inspection, occupational health and safety practices, and risk mitigation strategies.

Expert guidance on commissioning and handover, facility documentation, owner training, and maintenance for long-term sustainability.

Essential information on post-construction evaluation, continuous improvement, professional development, and knowledge management in construction management.

This book is not just a theoretical guide; it is a practical companion for construction professionals, project managers, and students looking to enhance their skills and achieve outstanding results.

The Art Of Construction Project And Business Management

This resource is a comprehensive and indispensable guide for professionals in the construction industry. The groundbreaking book unravels the intricacies of managing construction projects and businesses with precision, expertise, and a strategic approach.

Drawing on years of industry experience and the latest industry insights, the book takes you on a transformative journey through every stage of the construction project lifecycle. From project initiation to execution, control, and beyond, you gain a deep understanding of the key principles, best practices, and real-world challenges faced by construction project managers and business leaders.

The book goes beyond theoretical concepts, offering practical, actionable strategies that can be applied in your daily work. With meticulous attention to detail, it equips you with the skills, knowledge, and tools to achieve project success, drive business growth, and navigate the ever-evolving landscape of the construction industry.

Inside, explore essential topics such as project planning, scheduling, and risk management, ensuring your projects are well crafted for success. Learn how to navigate complex legal and regulatory frameworks while maintaining the highest ethical standards and corporate social responsibility.

Real-world case studies provide invaluable insights and lessons learned from notable construction projects. These practical examples showcase the application of best practices and demonstrate how to overcome common challenges, ensuring you're equipped to tackle any project with confidence.

A Practical Approach To Risk Management In Construction: Unlocking Project Success

Discover the essential guide to practical risk management in the construction industry. This book provides construction professionals with the knowledge and tools they need to navigate the complex world of risks and ensure project success. From identifying and analyzing risks to developing effective mitigation strategies, the book offers practical techniques and real-world examples that empower you to proactively manage risks and deliver exceptional projects.

By delving into the critical aspects of risk management, you'll be equipped with the skills to anticipate and address potential challenges throughout the project lifecycle. Learn how to incorporate risk assessment into project planning, engage stakeholders in risk identification, and implement robust risk response strategies. The book also explores the integration of sustainability considerations into risk management, showcasing the industry's evolving focus on environmentally responsible practices.

Engage with engaging case studies and lessons learned from real-life scenarios, gaining invaluable insights into the application of risk management in diverse construction projects. Emphasizing the importance of practical approaches, the book provides actionable strategies for addressing risks effectively, maximizing resource utilization, and fostering collaboration among project stakeholders.

Explore technology-driven risk analysis tools, such as data analytics and artificial intelligence, demonstrating how these innovations enhance risk management outcomes and decision-making.

The Dictionary Of Construction Terminologies: A Compendium Of Knowledge For Students, Academics, Practitioners, And House Owners

The dictionary of construction terminologies book is a comprehensive reference guide that provides definitions and explanations of the technical language and jargon used in the construction industry. It is an invaluable resource for professionals working in construction, as well as for students learning about the industry or for individuals looking to understand construction-related concepts better.

The book features a wide range of entries that cover various aspects of construction, including architecture, engineering, materials, equipment, and techniques. The book also provides clear and concise definitions of technical terms, written in easy-to-understand language. Terminologies are presented alphabetically to help readers find the descriptions they need quickly and easily. Whether you are a professional working in the field or interested in construction, this book is an essential tool to help you navigate the complex world of construction terminology with confidence and clarity.

Are you a student of construction, a house owner, an academic in the construction industry, or a practitioner that desires to acquire more knowledge about construction terms? If your answer to the preceding question is affirmative, this book may be one of the best investments you will ever make.